MW01321084

Urban Flood Risk Management

Like so many of the coastal cities in Southeast Asia (and other regions) established during European colonialism, there has been an ongoing challenge for decades dealing with the growing frequency and intensity of flooding. Jakarta's flood problems since the 1990s have been nothing less than monumental and the inability of the local and national governments to mitigate flooding in Jakarta is the most visible manifestation of fundamental water management deficiencies. This book offers a comprehensive and systematic historical assessment of Jakarta's water management practices from the colonial era through the early years of the Indonesian republic and Jakarta's emergence as a sprawling megacity.

This book draws upon a vast multidisciplinary literature and a wide array of government documents to unravel the complex history of water management that has led to approximately 40% of the city now lying below sea level.

This book will be a useful reference to those who research on topics such as urbanization in Southeast Asia, sustainable development, urban and planning history, environmental planning, issues of water management (and flooding), and the politics of planning and development.

Christopher Silver, PhD, FAICP, is Professor at the Department of Urban and Regional Planning, University of Florida. He is a four-time Fulbright Senior Scholar in Indonesia and holds honorary professorships at the University of Indonesia and the Institute of Technology, Bandung.

Routledge Research in Sustainable Planning and Development in Asia
Series Editor: Richard Hu

Urban Flood Risk Management
Looking at Jakarta
Christopher Silver

For more information about this series, please visit: www.routledge.com/Routledge-Research-in-Sustainable-Planning-and-Development-in-Asia/book-series/RRSPDA

Urban Flood Risk Management

Looking at Jakarta

Christopher Silver

LONDON AND NEW YORK

First published 2022
by Routledge
2 Park Square, Milton Park, Abingdon, Oxon OX14 4RN

and by Routledge
605 Third Avenue, New York, NY 10158

Routledge is an imprint of the Taylor & Francis Group, an Informa business

British Library Cataloguing-in-Publication Data
A catalogue record for this book is available from the British Library

Library of Congress Cataloging-in-Publication Data
Names: Silver, Christopher, 1951– author.
Title: Urban flood risk management : looking at
Jakarta / Christopher Silver.
Description: Abingdon, Oxonn ; New York, NY : Routledge, 2022. |
Series: Routledge Research in Sustainable Planning and Development in Asia |
Includes bibliographical references and index.
Identifiers: LCCN 2021021590 (print) | LCCN 2021021591 (ebook)
Subjects: LCSH: Flood control–Indonesia–Jakarta. | Urban runoff–
Indonesia–Jakarta–Management. | Storm sewers–Indonesia–Jakarta.
Classification: LCC TC513.I55 S48 2022 (print) |
LCC TC513.I55 (ebook) | DDC 627/.40959822–dc23
LC record available at https://lccn.loc.gov/2021021590
LC ebook record available at https://lccn.loc.gov/2021021591

ISBN: 978-0-367-77427-1 (hbk)
ISBN: 978-0-367-77430-1 (pbk)
ISBN: 978-1-003-17132-4 (ebk)

DOI: 10.4324/9781003171324

Typeset in ITC Galliard
by Newgen Publishing UK

Contents

Illustrations

Tables

Introduction
Rising waters

In January 1990, after having been living in Jakarta for just three months, the annual rains came with their usual vengeance. The "rainy season," as locals referred to it, begins typically in November or in December, but the heaviest downpours, the type that can last all day (rather than just dampening the late afternoon) seem to hit in January and February. According to the locals, these rains typically cause just some temporary inconveniences but are a part of the Jakarta experience. That year was different, however. The continuous rainfall shut down large sections of the capital city, hitting especially hard those "slum areas," as the *Jakarta Post* referred to them, in South, West, and East Jakarta. A child drowned when the Krukut River, passing through the Pondok Karya neighborhood in South Jakarta, overflowed its banks. The highest water recorded was roughly one meter in the Rawa Buaya area of West Jakarta, with roughly one-half of that depth reported in other low-lying areas throughout the city. North Jakarta neighborhoods, adjacent to Jakarta Bay, experienced flooding both from the inundated rivers and clogged canals unable to handle the rainfall, as well as the rising sea that pushed water back onto shoreline settlements.[1] Overall, however, residents treated the 1990 flood like other previous similar events as little more than a temporary inconvenience that created short-term traffic problems and shut down schools. As usual, the heaviest burdens were borne by the poorer residents largely confined to living in low-lying areas often adjacent to the rivers and canals. Even where we lived on the relatively high ground of the Kebayoran Baru neighborhood in South Jakarta, for several days many streets became virtual rivers. In the days following the flood and as the waters receded, public officials indicated that there should be more attention to improving the drainage system, perhaps dredging the waterways to increase the flow. These statements were the same annual pronouncements during the rainy season that made for several decades but rarely followed up by actions.

As it turned out, the 1990 flood was a prelude to a series of inundation episodes of substantially greater magnitude over the next two decades. Floods that were far more devastating occurred in 1995, 1996, 2002, 2004, 2007, 2013, 2015, and 2020.[2] The three floods that occurred in succession between October 1995 and February 1996 resulted in 20 deaths and displaced temporarily nearly half a million Jakarta residents. Again, I was living in South Jakarta, and in this instance,

DOI: 10.4324/9781003171324-1

a family we knew living in a nearby high-end neighborhood constructed inappropriately in a lowland site adjacent to Kebayoran Baru had to be rescued from their second-story windows by the Indonesian Marines in a boat cruising by in water nearly four meters in depth. It was definitely worse than the 1990 flood, but not enough to trigger sustained post-flood interventions.

The flood in 2002, identified as the worst in the city's history, displaced approximately 300,000 people and claimed many more lives.[3] It affected a substantially larger portion of the city than the 1996 flood, including an unusually large number of Jakarta's higher income areas as well as the usual victims, low-income communities living in the flood zones. A luxury community in East Jakarta, Cipinang Indah, endured inundation that at its high mark recorded two meters of floodwaters. The high waters of the 2002 flood also led to a significant conflict between the upper-income ethnic Chinese Kelapa Gading community and the adjacent Sunter area. As the waters rose in Sunter River, the city closed a floodgate to protect the Astra car factory in Sunter owned by a prominent Jakarta business group (which included the son of Indonesia's former president). Closing the floodgate directed the high waters into the Kelapa Gading enclave, turning the whole community into a virtual lake, with rafts and boats the only functioning vehicles. Kelapa Gading remained submerged for nearly one week.[4] By February 2, when the rains had still not ceased, the national government finally declared an emergency. By then, the flood had hit many previously unaffected portions of the capital, including high water surrounding the presidential palace, and traffic shutdown around the central business areas in the old city (Kota) as well as businesses along Jakarta's main commercial corridors. All four thoroughfares, M.H. Thamrin, Jendral Sudirman, Gatot Subroto, and H.R. Rasuna Said, were flooded, with "all offices including banks ... closed."[5] The devastation brought on by the 2002 flood exacerbated the economic problems Jakarta already were dealing with in the aftermath of the 1998 Asian financial crisis, an economic crisis severe enough to displace the five-term president Suharto and trigger the demise of the indomitable Golkar political party. Despite the devastation, it did not take long to forget what had happened in 2002 and what Jakarta needed as remedies to prevent future disasters. Within a year, all of the post-disaster pronouncements of flood mitigation actions that the city intended to implement had disappeared from the policy agenda.

Not so five years later when next "historic flood" occurred. The flood conditions of the 2007 flood greatly surpassed the devastation of the 2002 flood and permanently positioned flood mitigation at the top of Jakarta's (and the nation's) policy agenda. The 2007 flood dwarfed all four of the previous flood in every respect. According to a report prepared by the Asian Disaster Preparedness Center (ADPC) in Bangkok, "the floods of February 2007 were the worst in the history of Jakarta, covering almost 60% of the urban area." Estimates of the number of residents made homeless over the worst three days of the flooding ranged between 240,000 and 512,170, with a final tally of 57 persons killed either through drowning or electrocution, with thousands more suffering from health effects of having to wade through several meters of polluted waters to escape the floods.[6] The financial losses from the 2007 flood reached an estimated 8.8 trillion rupiah (or US$871.2 million). The only part of the city largely spared flooding in 2007 flood was the government and business centers in Central Jakarta. Just one portion of Central Jakarta experienced flooding in excess of 60 centimeters.[7] Kampungs packed along the city's rivers were not so fortunate, however. According to one report, Kampung Melayu situated alongside the Ciliwung River, the largest of the 13 rivers running through Jakarta, and frequently the epicenter of the floodwaters raging down from the mountains, experienced six meters of flooding. Kampung Melayu resided adjacent to a section of the river where years of neglected river maintenance reduced the depth of the channel to just a couple of meters, thereby limiting its capacity to handle the volume of water.[8]

The devastation brought on by the 2007 flood permanently elevated flood risk management to a top priority for Jakarta's political leadership and generated an unprecedented flurry of assessment to identify the causes of the increasingly disastrous floods and to devise and implement necessary mitigation strategies. While the flood assessments and proposals to correct the problems came quickly, implementation of flood mitigation projects was neither quick enough nor sufficient to prevent similar catastrophic floods over the next decade. The flood in January 2013 caused by the collapse of a section of the West Flood Canal, a structure built in the 1920s under the Dutch regime, led to 47 deaths and mass flooding in the previously unaffected central business district. Images on the front pages of local newspapers captured the flooded and impassable traffic circle in front of Plaza Indonesia where Jalan Thamrin and Jalan Sudirman connect and showed readers that no areas were safe. Another poignant image in the *Jakarta Post* showed President Susilo Bambang Yudhoyono standing inside the presidential palace with his pant legs rolled up to keep dry from the floodwaters that had made it inside. The February 2015 flood, a consequence of 18 inches of rain falling within a 24-hour period, swept through 307 neighborhoods in West, North and Central Jakarta, shut down the city's bus rapid transit system, and again sent water into the presidential palace located on usually high and dry land surrounding Merdeka Square.[9]

On New Year's Eve of 2020, heavy and sustained rains set a new one-day record in terms of water volume and triggered another round of mass flooding, this time in sections of two bordering cities, Bekasi and Tangerang, as well as landslides in the suburban locale of Depok through which the Ciliwung flowed.[10] Within the Jakarta administrative borders, the community of Bidara Cina also situated alongside the Ciliwung experienced water up to the rooftops. In an unprecedented move, residential and commercial property owners filed a lawsuit against Jakarta's governor, Anies Baswedan, claiming the government had failed to protect them properly and demanding compensation for their losses. Within a couple of weeks, however, coverage of the 2020 flood issue disappeared from the media. As we will see, outrage followed by acceptance was not an uncommon response to even the worst floods in a city long used to dealing with flooding as an unpleasant fact. Although flood mitigation remained a top priority for local administration continuously after the historic 2007 flood, none of the proposed or implemented interventions proved to be the solution that the city needed.

Perhaps the lack of post-flood coverage and opining on the causes of the 2020 flood by the media reflected the reality that the causes and proffered solutions to Jakarta's flooding were now widely known given over two decades of lengthy post-flood assessments. Harkening back to the 1990 and 1995/1996 events, the initial post-flood analyses tended to emphasize the most common explanations, namely too much rain too quickly and the inability of the rivers to accommodate the volume of water, especially what was coming into the city from upstream locations. Another factor cited was massive urbanization of Jakarta's periphery. These developments reduced the green areas needed to absorb the rain and to control runoff. Added to the widely understood problems of managing the effects of heavy rains was evidence, widely disseminated in the aftermath of the 2007 flood, that at least 40% of Jakarta's land had sunk below sea level, especially portion of North and Central Jakarta. Land subsidence became the new major causal factor affecting the elevated scale of flooding and a problem that would require complex interventions to deal with.[11] It prompted a new sense of urgency to flood management since the international media proclaimed that Jakarta, owing to land subsidence, had become one of the fastest sinking cities in the world.

A better explanation why the devastation of the 2020 flood failed to prompt sustained news coverage is that Jakarta's future as the nation's capital was no longer certain. Early in 2019, immediately following the victory for a second term as president, Joko Widodo announced that the government planned to move the nation's capital from Jakarta to a more centralized (albeit remote) location in Indonesia. One of the key factors influencing the decision of creating a new capital was the recurring flooding and how this interfered with government operations. So if Jakarta was no longer going to be the capital city, how important was it to invest in flood mitigation? The conversations in the media shifted from floodproofing the current capital to creating a new "green" and "sustainable" city unburdened with the polluted, flood-prone rivers that characterized the environment of Jakarta in 2020.

Policy paralysis, or perhaps policy failure, accounted for the relatively tepid responses to another major flood in 2020 and the decision to relocate the capital. Despite over two decades of flood mitigation interventions in response to the string of historic flooding events since the 1990s, responses that involved significant public funding to address the problem, Jakarta remained a city perpetually "under water." Why? As Pelling and Blackburn suggest, "Jakarta's response to the interrelated problems of water management and supply, pollution, waste disposal and flood risks – and the possible impact of climate change on this nexus of issues – remains arguably piecemeal."[12] The incidence of major floods attested to the failure of effective water management throughout Jakarta's sprawling megacity region. In this sense, flooding was just one manifestation of a more complex set of problems within the water management system. Given the wealth of studies that pointed to the need for comprehensive and integrated urban water management, the piecemeal approach by Jakarta, usually attributed by local leadership to the lack of sufficient funding, was incapable of producing the desired outcomes. Not just a lack of sufficient funding but also a lack of willingness to take necessary steps to confront the dysfunctional water management system must also be considered.

Why eight major floods in 25 years?

Recommendations on how to address Jakarta's annual flood conditions were nothing new, appearing in the local and national planning documents beginning in the 1950s. In the aftermath of the devastating floods of 2002 and 2007, thorough analyses of the causes of Jakarta's flooding generated new understanding of the complexities of the problem and the challenges of coming up with effective mitigation actions. Scientific and technical studies motivated by dissatisfaction with the piecemeal interventions cited by Pelling and Blackburn identified the complexity of factors contributing to flooding and a daunting array of interventions needed to mitigate it. Commonly cited by experts and pundits alike was the volume of rainfall within such a short time span. This was nothing new and clearly a contributing factor in the January 2020 flood. Since the rains fed the multiple rivers and canals that lace through all sections of metropolitan Jakarta, heavy rains could affect simultaneously a vast area. By the late 20th century, the built-up areas of Jakarta had spread out to encompass the watersheds of all 13 rivers that flowed from the southern highlands through the metropolis. During the rainy season, runoff from densely populated portions of the city, supplemented by runoff in the rapidly developing periphery (which was where some of the worst 2020 flooding occurred), added to the floodwater volume.

During the first three centuries of the city's development as the Dutch colonial administrative center named Batavia (discussed in detail in Chapter 2), all but two of Jakarta's rivers principally irrigated agricultural lands that surrounded the compact coastal settlement. These agricultural lands absorbed much of the annual rainfall, thereby generating far less runoff into the waterways than was the case once the impervious surfaces of Jakarta's megaurbanization covered the land. Although there had been a continuous problem of silting at the mouth of the Cidane and Ciliwung rivers that passed through the original settlement in the 17th and 18th centuries and sometimes caused flooding, the volume of rainfall was never the serious problem that it became beginning in the late 20th century. In fact, a continual and more serious challenge during the early years of the city's existence was not enough water to support even a modest-sized settlement and its agriculture, especially during the "dry season."

Planting a city in a river delta on land just a few meters above sea level, fed by multiple rivers and with significant annual rainfall, virtually guaranteed that flooding would occur on some regular basis. And it did. The history of Batavia/Jakarta over four centuries of development includes evidence of every recognized type of flooding. These included river flooding, coastal floods (especially in the settled sections located on Jakarta's north coast), flash floods which occurred during tidal episodes or when flood protection infrastructure failed; stagnant floods during extreme rainfall when drainage systems were blocked; and canal floods when breaks in canal structure or the lack of pumping capacity leads to overflows onto adjacent land. In virtually all of these situations, heavy rain over a short time span is a credible explanation. Jakarta received plentiful rain in heavy dosages.

Beyond the meteorological forces generating tidal surges and excessive rainfall, anthropogenic factors, that is, how the city itself developed in relation to its hydrological systems over the past three centuries, contributed to the location, duration, and impacts of flooding. One dimension of this was how urbanization removed green areas critical to managing stormwater flows. From the earliest settlement, indigenous communities resided alongside the rivers and streams, as well as in surrounding wetlands, but without causing any disruption of these natural processes. These communities relied on the rivers for sustenance and respected them for their value. Throughout the city's first two centuries, development along the canals and rivers typically had the advantage of high ground made possible by compiling the materials extracted to create the canals and to dredge the rivers. So flooding rarely was a problem. Indigenous settlements in low-lying and flood-prone areas built their simple houses as elevated structures in order not to hinder the natural flow of the water.

The controlled development processes that characterized city expansion all the way through the mid-twentieth century ceased when Jakarta became the capital city of the new Indonesian republic and experienced rapid growth. Inadequate enforcement of regulations on development contained within the city's planning schemes allowed dense development to push up to the edges of the rivers, streams, and canals and to pave over green areas designated as water catchment zones. This increased stormwater flows into rivers, abetted by road and community infrastructure that purposely sent the surface waters into these water bodies. Ongoing deforestation deep into the urban periphery was another outcome of unregulated urbanization that exacerbated surface runoff. The residential settlements allowed to crowd along the riverbanks in the city and beyond contributed to erosion that over time resulted in a notable reduction in flow capacity of the waterways. This meant they were less able to handle the annual rains that they traditionally managed so well. In colonial Batavia, building alongside the rivers was not a problem because when erosion or debris from these dwellings and businesses built up, regular maintenance removed the problem materials to keep them clear for commerce. These same waterways provided residents with water for consumption and accommodated household waste that the rains and routine maintenance carried into the sea. Maintaining the city's footprint on a narrow slice of high ground extending southward between the rivers and canals enabled residents to keep relatively dry in this water city.

As the city population expanded in the 19th and 20th centuries, the surface water resources underwent significant degradation. At the same time, the city ceased to maintain regularly the rivers and canals once their role in local transportation and commerce declined. Indigenous communities continued to rely on them for drinking, bathing, washing clothes, and as waste receptacles, but the growing European population turned to groundwater and later on added sanitation systems to support their domestic needs. The physical expansion of the urbanized area resulting from population growth after 1900 pushed new settlements into lowland areas where flooding naturally occurred even without heavy rains. Independence in 1950 prompted mass migration to Jakarta, and this resulted in unregulated informal settlements forced to occupy undeveloped spaces alongside all of the city's rivers. City officials gave little notice to any potential impacts on river water quality by allowing settlements to cluster along its waterways. These were not lands desired by new formal developments and were alternative housing options provided by the government or the private sector to meet the growing needs the masses pouring into the city. As the quantity and quality of surface water diminished, the use of groundwater accelerated. By the early 1970s, groundwater became the primary water source for a city that now exceeded 4.5 million. Some secured water through the public water company but more commonly relied upon private wells. In the 1970s, analysts first noted evidence of land subsidence in areas of Jakarta, a factor later acknowledged as contributing to intensified flooding. Land subsidence was most evident in North Jakarta along the coast of the Java Sea. Although recognized as a problem that needed to be addressed, there were no connections made between the sinking land and the increased reliance on groundwater. Because land subsidence occurred predominantly along the city's north coast, a floodwall to protect businesses and settlements on the depressed lands from rising seas and storm surge to mitigate flooding was undertaken. Another impact of land subsidence was on the flow of rivers that passed through areas of subsidence. As the riverbeds sank, they could no longer empty directly into the sea. This meant adding new pumping technology to carry waters over the floodwall. During the rainy season when river water volumes increased rapidly, the pumps needed to work overtime to carry the fast-rising waters over the floodwall to protect settlements on the landside from flooding. If the pumps failed, which occurred at times, the resulting flooding was so much worse.[13]

Extensive research on the causes of Jakarta's land subsidence determined that it resulted from the city's overreliance on groundwater, a condition owing to a combination of factors. One factor causing this overreliance was the polluted condition of local surface waters. Surface water degradation was not new in Jakarta, but the heightened levels of pollution meant that even with treatment, it could not be used either for domestic or industrial consumption. Unregulated urban development along the city's waterways, especially the lack of sewerage service for communities and businesses in the city and upstream, directly contributed to the elevated pollution levels. As a result, Jakarta needed to look outside its environs for surface water clean enough to supply its piped water system. The principal source was a reservoir located 70 kilometers east of the city. Constructed in the mid-1960s, the Jatiluhur reservoir got its water from the Citarum, a river that originates in the highlands south of Bandung. Although the largest reservoir in Indonesia, Jatiluhur, served more than the main source of Jakarta's surface water supply. It supported hydroelectric power generation and provided the major source of water for irrigating regional agriculture. So just a small percentage of its water makes its way to Jakarta through a system of canals. As a result, the major source of water for the Jakarta metropolitan region is its deep and shallow aquifers. Batavia began tapping groundwater for city uses as early as the mid-19th century. The volume of extraction was limited initially because surface water remained in wide use. So long as the city retained ample green space, at least some of the shallow aquifer recharged during the annual heavy rains. Throughout the 20th century, as surface water use declined to support the "formal" sector, subsurface water consumption accelerated. As previously noted, planning documents first mention land subsidence in the 1970s but not directly connected to groundwater extraction. Not until the disaster assessments following the 2007 flood did analysts confirm the connection between land subsidence and overreliance on groundwater extraction and make the connect to the rising seriousness of flooding. The extent of the land subsidence in North Jakarta, accompanied by high tides and the rising waters in rivers, explained why the 2007 flood set the all-time record for the depth and breadth of flooding throughout the urbanized area.[14]

Jakarta in regional context

As Jakarta's leadership tried to understand why the city's flood episodes escalated in intensity beginning in the mid-1990s, they just needed to look at what was happening in the other rapidly growing neighboring megacities in Asia for the answers. Similarly situated cities throughout Asia experienced intensified flooding at the same time, typically owing to many of the same dynamics underway in Jakarta. Owing to European colonial interventions in an era of water-based transport, most of Asia's emerging megacities began as trading posts situated on coastal plains or upstream on navigable rivers that made them vulnerable to flooding. All experienced rapid growth often with insufficient concern for the impact of urbanization on the environment. From India through Southeast Asia to East Asia, there were no fewer than 30 cities with a population of 10 million or more, and eleven of these boast populations of greater than 20 million within their extended megapolitan region. Of those large enough to qualify as megacities, Manila, Mumbai, Karachi, and Jakarta sit in a coastal plain, while Shanghai, Guangzhou, and Kolkata are located in a delta and all experienced flooding in recent decades. Several other Asian megacities located upstream, such as Bangkok, Dhaka, and Ho Chi Minh City, also experienced severe flooding similar to the coastal cities.[15]

In a study of Ho Chi Minh City's vulnerability to the flood-related effects of climate change, Storch and Downes (2011) emphasize that "all studies identify Southeast Asia as the flood prone region with the greatest need for urgent policy measures" based upon factors including the incidence of storm surge and land subsidence.[16] Scientists first detected subsidence in Ho Chi Minh City in 2003. Geologists determined that weak soil was one factor. More important was the increase in impermeable surfaces because of rapid urbanization coupled with the estimated 200,000 boreholes constructed to extract groundwater to serve the growing city. Although Ho Chi Minh treated available surface water, groundwater was cheaper and more reliable than what was provided by the local water company. According to Nguyen (2016), better control of urbanization and regulation of tapping into groundwater was the only way to slow land subsidence and mitigate the flooding. The absence of adequate regulation of urban development allowed development to invade Ho Chi Minh's most environmentally sensitive lands and to trigger greater flooding.[17]

Like in Ho Chi Minh City, intensive flooding is the reality of urban life in low-lying waterfront megacities located upstream. A case in point is in Bangladesh where the devastating flooding in September 2017 covered one third of the entire country. Its greatest impact was in Dhaka, the megacity region of more than 17 million inhabitants, regarded "among the most climate-vulnerable megacities in the world."[18] Modern Dhaka, one of the world's densest cities, sprawls across a wetland formed by four major rivers (Buriganga, Turag, Thong Khal, and Balu-Shitalakha) assuring the existence of water everywhere in the city region, and much like Jakarta. Between 1954 and 2009, Dhaka experienced ten major floods, some a result of upstream rains that exceeded the rivers' capacities but also because of excessive rainfall within the city and the lack of protections against rising water in low-lying areas. The 1998 flood inundated the city for 65 days, hitting hardest its poorest neighborhoods. According to a 2015 World Bank study, the increased flooding had many causes, but the primary cause seemed to be the transformation of wetlands to accommodate new urban development that reduced the capacity of natural protections against flooding. In addition, overreliance on groundwater for urban consumption had reduced the deep aquifer by approximately three meters per year and the resulting land subsidence exacerbated flood conditions. Given the city's position within an extended floodplain region, the World Bank study recommended that its risk management strategy needed to include green defenses and an ecosystem-based strategy rather than focusing just on hard infrastructure.[19] Again in July 2020, Dhaka experienced the worst impacts of another nationwide flood event that submerged approximately 37% of the city.[20] Other Asian megacities located on coastal plains have experienced equally devastating flood conditions, including Mumbai, Karachi, Kolkata, Yangon, Manila, Shanghai, and Osaka.[21] Cities situated upstream such as Seoul, Chennai, Shenzhen, Bangkok, and Guangzhou are just as vulnerable to biophysical hazards as those in the coastal regions.[22]

Flood events make front-page news but typically are just the most visible manifestation of far more complex environmental management challenges in rapidly growing Asian cities. Significant deficits in drinking, sanitary, and stormwater infrastructure in many of these cities contribute to, and often accelerates, flooding. Limited use of surface water for urban consumption seems to be a universal situation that leads to overextraction of groundwater and evidence of land subsidence.[23] Fuchs underscores this relationship. As he states, "many coastal megacities in Asia are built on deltas where significant sinking is occurring due to soil compaction or groundwater withdrawal for household and industrial purposes."[24] As cities sink and sea-level rise and storm surge increase, the result is deeper and more expansive flooding. Analysts linked increased flooding in Bangkok to global warming and sea-level rise in the Gulf of Thailand coupled with the problem of the upstream city sinking at an annually unsustainable rate. A 2006 study of land subsidence in Bangkok confirmed that the city had been sinking for the previous 35 years, and this owed largely to the effects of deep well pumping. Land subsidence accounted for the intensity of the city's 1995 flood. In its aftermath, Bangkok took measures to limit groundwater use and to mitigate future flooding. To ensure greater reliance on surface water, the government imposed higher pricing of groundwater and instituted strict enforcement of prohibitions on new wells.[25] The cases of Ho Chi Minh City, Dhaka, and Bangkok demonstrate the challenges of planning, water management, and flood risk management throughout the region. Excessive development on environmentally sensitive lands, overextraction of groundwater leading to land subsidence, and inadequacy of environmental infrastructure all contributed to the catastrophic flooding experienced in Asian cities. As the Bangkok case demonstrates, reducing land subsidence as a mitigation strategy for flooding requires interventions that involve the city's entire water infrastructure. As in Bangkok, Jakarta's flood risk challenges reflected the city's overall water management system.

Indonesia's fastest sinking cities

Among Asia's coastal megacities, the most significant cases of land subsidence and associated flooding occur in Indonesian cities along the north coast of Java, including not just Jakarta but also Semarang in Central Java and the country's second largest city, Surabaya, in East Java. Areas of Jakarta lying along the coast of the Java Sea sank up to four meters below sea level over the past one-half century. Semarang and Surabaya experienced land subsidence of between two and three meters along their coastal areas. The implications of this for the natural drainage systems are significant. It means that the none of the rivers in Jakarta feed directly into the sea (except where they are diverted into flood canals) and require pumps to carry floodwaters over a seawall that the government built to protect areas now situated below sea level. In the case of Semarang, the city constructed a polder within the city to accommodate the high waters that cannot naturally flow to the sea. Without the pumping system installed in Jakarta, which sometimes breaks down because debris clogs the system or there is a power failure when it is most needed, river waters overflow into adjacent settlements. The seawall itself is under stress since the land on which it sits experiences subsidence. To keep pace with the sinking land, Jakarta must add continuously to the coastal seawall height.[26]

Recognizing the problem of land subsidence and its connection to how extensively the city relies on groundwater clarified that the challenge of managing the substantial annual rains and resolving the flood problem involved more than building barriers to keep water under control. It brought into discussion of flood mitigation strategies the role of anthropogenic influences, especially urbanization and spatial expansion of the built environment, on the city's water resources both below and above the ground. Development impacts on the city's waterways and its water resources need to be centermost in governance processes. Poorly regulated urban development contributes to the loss of permeable open lands and destruction of the wetlands that are essential components of flood risk management.[27] The absence of regulations to prevent the construction of formal and informal settlements in floodplains alongside rivers represented another assault on the natural processes of the environment in Jakarta and other Asian cities.[28] Allowing riverfront settlements to literally spill over the shores throughout the urbanized area contributed to erosion that reduced the flow capacity of the rivers and accelerated stormwater runoff beyond what the rivers traditionally handled. The government's general indifference to the impacts of urbanization on the environment was evident in the earliest plans prepared to guide Jakarta's development as the new capital city. Although the master plans prepared in the late 1950s called for preservation of a "greenbelt" to contain urbanization, to absorb rainfall, to support aquifer recharge, and to help minimize the problems of runoff, no mechanisms existed to realize the ideals expressed in the plans. Jakarta's leadership routinely chose growth over environmental preservation. Even enforcement of a greenbelt was insufficient to provide the protections necessary to sustain the quality of its rivers and subsurface water resources. The greenbelt concept was appropriate to a moderate-sized city but not a rapidly growing place like Jakarta. Environmental Minister Rachmat Witoelar admitted as much in the aftermath of the 2007 flood. Despite greening strategies in Jakarta's spatial plans, he noted, "the elimination of water catchment areas" was a leading factor in the 2007 flood, and this represented the decision to build rather than to preserve. It is telling that there is no evidence that during Witoelar's term as Environmental Minister he initiated actions to alter the prevailing practices.[29]

Throughout the formative period of Jakarta's transformation into a megacity between the 1960s and the 1990s, residential, commercial, industrial, and institutional development voraciously consumed green spaces within and beyond the city proper. Simultaneously mass migration to Jakarta from all corners of Indonesia led to a building frenzy wherever there was unclaimed space, especially alongside its rivers, canals, and lakes. These self-built communities relied upon communal taps or private vendors for clean water. More important, they lacked access to sanitary infrastructure and had to rely upon the adjacent rivers and canals to deposit their household waste. Since there was no intention on the part of the government to extend full clean water and sanitarian services to these settlements, residents had no other choice but to make use of the rivers, streams, lakes, and canals. As long as the rivers and canals primarily serviced the overcrowded settlements that lined their shores, regular maintenance was not a priority of the government. With the lack of regular maintenance, solid waste and silt built up in the rivers. Removal of the debris was less important except and when the rains came and when the waste clearly contributed to flooding. In general, what once had been scenic waterways (as historic images of Batavia reveal) are now treated as convenient places to dispose of trash, to funnel stormwater, and in several areas to provide water for irrigating agriculture and golf courses. Reduced in size and grandeur, the murky brown still waters of Jakarta's rivers, typically dotted with floating waste, much of it coming from upstream sources, now appeared as open sewers and not the asset that they once had been. Only during the rainy season, when the volume of water cascading down from the mountains increases, do some of them regain the physical stature of the once mighty rivers they once were. Yet in this state, they are feared and scorned as the sources of destructive flooding.

It is very telling that Jakarta makes no effort to celebrate its rivers or to create public access to its watery environs. There are no river walks, nor any signage directing residents or visitors to parklike settings along its waterways. The one exception is a linear park constructed alongside the Pluit Reservoir along the north coast where the city pumps floodwaters into the sea during the rainy season. Apart from this recent addition of green space along a water body, and a small restored segment of the former Kali Besar (Grand River) in the old city (*Kota*) that is literally a disconnected pool of murky water, one would never know that there are 13 rivers and its tributaries flowing through the city or that this river system was the most important reason for creating the city in the first place. Except for the engineered flood canals that run alongside several major streets, Jakarta's natural waterways are invisible, inaccessible, polluted, and not a feature that the city cares to acknowledge.

For the informal communities that line their shores, the rivers are a key source of support. Some informal settlers literally built their structures into the river itself to gain more livable space. This explains why such a high proportion of the flood victims involve residents of the kampungs located within the watershed. This close physical and functional association with the rivers contributes to the widely propagated view that the informal settlements are not just flood victims but contributors to the excessive flooding because of how they use the waterways. Recent research shifts the focus to other actors, however. As Goh (2019) contends, the lack of regulation to prevent large-scale dumping by commercial and agricultural concerns, coupled with uncontrolled urban development that decreases ground permeability and groundwater infiltration, is a more relevant factor explaining the highly polluted conditions of the rivers and why they get clogged and flood during extreme rain events.[30] A study conducted on the plastic waste in Ciliwung River determined that the highest proportion of micro plastics detected at various points seemed to come from plastic bags and food wrappings discarded there.[31] Whether it is riverfront community residents or the general community at large, the use of the Ciliwung, as with other Jakarta rivers, largely as waste receptacles rather than respected natural resources is undeniable. Rather than regarding riverfront kampung residents as willful polluters, Leitner, Colven, and Sheppard point to the decision by the government to refuse assistance to secure access to appropriate sanitation services. Consequently, informal settlement residents have no other alternative to using the rivers and canals to manage their waste.[32] These informal settlements do not need to do this secretly since along all of the waterways one can see pipes projecting out of dwellings to dispose of wastewater into the rivers. How to provide alternatives to this common practice and how to reduce the overreliance on groundwater are the real problems to be resolved, and for which the government has not provided answers. The absence of civic standing experienced by such a large segment of Jakarta's population is substantiated by the denial of connections to wastewater and solid waste services available to other residents. This lack of civic standing is a more appropriate way to understand the linkage between the informal settlements and Jakarta's flooding crisis. If the needs of the informal settlements received the accommodation afforded to others, what would be the condition of Jakarta's rivers or the problems of flooding? This is worth considering.

Finding solutions

The sheer scale of flooding in 2002 and 2007 in Jakarta raised concerns about the consequences of unregulated urbanization and the capacity of public officials to respond effectively to flood challenges and to manage the city's water resources. World Bank environmental experts agreed that those who blamed Jakarta's flood problems largely on the actions of informal settlements missed the bigger problems. Insufficient maintenance and improper operation of flood control systems was high on the list. There was concern about solid waste services but particularly the practice of allowed large-scale commercial dumping in the rivers. They agreed with the view that land subsidence caused by excessive groundwater extraction and the inadequacy of aquifer recharge accounted for the increased intensity of Jakarta's recent floods.[33] One needed change was to do proper maintenance, namely, to dredge regularly the city's rivers and canals to enable them to handle heavy rainfall and runoff. Upgrading and extending the city's flood control infrastructure coupled with enforced land-use regulations to prevent development in flood-prone area was another recommendation. Extending the environmental infrastructure to reduce the routine use of the rivers for household and commercial refuse disposal was the big-ticket item and necessary to combat the pollution problem. Finally, the World Bank experts advocated constructing additional reservoirs or polders to potentially add more surface water but also to manage high-water episodes. Together, these actions constituted a flood management strategy that the experts advised Jakarta's leadership to embrace.

One strategy not touched upon by the World Bank analysts involved river restoration to utilize the capacities of the natural environment to protect against flooding. There were already in the global discourse on flood risk management suggestions on how "soft" infrastructure could be an effective form of mitigation. In the opening speech of the Third International Symposium on Flood Defense in 2005 in Nijmegen, Netherland, Melanie Schultz van Haegen, Netherland's State Secretary for Transport, Public Works and Water Management, noted that between 2003 and 2005 alone, there were 600 floods worldwide, claiming 19,000 lives and resulting in approximately US$25 billion in damages. In Dutch society, she noted, there is a love–hate relationship with water since it serves as the engine of the nation's economy but at the same time poses a continuous threat. Yet through a "system approach" to defend against floods, the Netherlands claimed an impressive record of virtually no significant inundations since the early 20th century. As Schultz van Haegen noted, "the last time flooding with dike failures in the catchment area of the major rivers in the Netherlands occurred" was in 1926. In 1995, 250,000 people had to evacuate but the dikes held.[34]

The year 1926 is significant for Jakarta as well. In that year, the city (when Jakarta was still the Dutch East Indies colonial administrative center of Batavia) completed the construction of the first segment of Jakarta's current system of flood control. The flood canal diverted water from several of the city's rivers around the settlement areas and conveyed it safely to the Java Sea. The colonial government never completed the flood canal system. Soon after World War II and the onset of Indonesia's independence, Jakarta's development pushed well beyond the area served by the existing flood canal system. Spatial expansion of Jakarta accelerated rapidly from the 1970s onward but without any major additions to the flood canal system through the end of the century. The lack of a completed flood canal system was not the main problem if one follows the new logic of the Dutch system. As Schultz van Haegen observed, the global challenge of properly managing to prevent floods resides not so much in the technology or engineering but in the mechanisms of guiding urbanization. "Governments make insufficient allowance in their spatial plans for the risks of floods. The short-term profit for new residential neighborhoods or industrial parks often prevails over long-term flood defence."[35] In the case of Jakarta, there was no protections of the lands along its rivers, and this lack of environmental protection rather than more flood infrastructure was the problem.

In the Netherlands, a fundamental policy change occurred in 2000 when the government decided to cease raising the height of its coastal dikes to counter future sea rise in favor of broadening the area where river waters could spread out to accommodate the larger flow during high-water episodes. This required tight controls on where to allow urbanization to take place as well as the density of development allowed near vulnerable places. In other words, the planning system needed to control urbanization, not the river. This approach, known as "Room for the River," focused on development along the rivers and wetlands and allowed river water space to go where it naturally flows during periods of heavy inundation without affecting residents or businesses.

Over the past 70 years of unprecedented urban expansion, the rivers that feed the delta region of Jakarta lost much of the space that allowed them to flow where they naturally did in the past. In direct contravention of the "room for the river" strategy, Jakarta's urbanization process constrained rather than accommodated the needs of its rivers. According to Pichel (2006), the "narrowed natural flow of rivers" was the most significant factor in the increasingly worse situation regarding flooding in Jakarta since the 1950s.[36] Beyond periodic calls for more effective flood control efforts, neither the system of water management nor the urbanization processes affecting Jakarta's waterways adequately addressed this problem. There seemed to be no connection between inadequate planning and floods, nor any enduring political consequences associated with the devastating costs of so much of the city going underwater.

The 2007 flood changed the political landscape. The 2007 flood had an immediate and sustained political impact, calling into question Jakarta government's effectiveness and responsiveness to a wide array of water management challenges. It triggered unprecedented and sustained pressures to confront the fundamental conditions associated with flooding. It led to calling in the Dutch engineers to advise the city on strategies. Flood risk management became one of the top policy priorities of the Jakarta's first popularly elected governor, Fauzi Bowo, in 2007.[37] Beginning under the Bowo governorship, and continuing with his three successors, Joko Widodo ("Jokowi"), Basuki Tjahaja Purnama ("Ahok"), and current governor Anies Baswedan (elected in 2017), flood management presented the great unresolved challenge. All of these administrations experienced the full complement of water resource challenges. These included flood protection, cleaning polluted rivers, expanding access to clean water, protecting critical waterways from urbanization pressures, and addressing the overuse of groundwater and the resulting land subsidence. The elevation of flood mitigation to the top spot of the local policy agenda recalled the governance priorities that had guided the city's development at strategic junctures since the 17th century, a time when water resources management was such a fundamental concern for the survival of the community.

The failure to find effective responses to the extensive flooding in Jakarta over the past decades prompted questions about whether the megacity could continue to remain the heartbeat of the nation. This engendered two very contrasting responses. One was, as noted above, to consider relocating the capital city to a less vulnerable place.[38] The other was to reconstruct Jakarta into the "water city" that had been its original purpose but to do so in a modern 21st-century form. The Jakarta Coastal Defense Strategy (JCDS), conceived by Dutch engineering consultants as a flood mitigation project to safeguard the existing megacity footprint, was nothing less than the conception for this new "water city." By constructing this new city on reclaimed land in the sea, with a reservoir (polder) located between the existing north coast and the new city to manage floodwaters, the JCDS would protect "sinking" Jakarta from the sea, stop inundation of the city's lowlands, and provide abundant new space to expand the city. The JCDS would restore the historic waterfront as the focal point of a floodproofed city. Like earlier waterfront schemes, the JCDS drew opposition from environmentalists and the local communities who recognized that this plan did not have their primary interests in mind even through it might protect against flooding. It was, as with previous plans for waterfront redevelopment in Jakarta, targeted to support the development community and its upper-income market, would generate displacement of low-income communities that had resided there for decades, and cause untold environmental harm to these coastal areas. As the preliminary sketches indicated, the glittering towers and marinas of the proposed waterfront city served a segment of the Jakarta community that did not include the existing low-income waterfront residents.

The fierce resistance to reclamation of the Jakarta waterfront, despite its promises to provide some long-term relief from flooding, brought into clear focus the contested politics of water management in Jakarta. Goh (2019) describes flood management in Jakarta as an example of what she terms "hydropolitics" whereby various stakeholders brought to governance differing views on planning, implementation, and maintenance of infrastructure systems, such as canals, gates, and catchments, as well as the planning and control of development and open space.[39] The political contestation inherent in Jakarta's water management practices involved groups of actors with radically different perspectives. Those government officials responsible for operation of the systems (flood control, waste management, provision of clean water, and pollution control) favored schemes that potentially reduced political fallout from flood events and that demonstrated action to resolve the problem. A second set of stakeholders includes residents affected directly and indirectly by the way the flood risk management systems function because they are either beneficiaries of effective response or victims of failures. A third set of stakeholders are those who have been excluded from key operations within the water management system, namely the majority poor population and, as previously noted, those treated as a source of Jakarta's water management deficits whether it comes to contributing to flooding or the degradation of the rivers, canals, and north coast along which they reside. A fourth group of stakeholders, the development community, favors a water management system that functions to increase opportunities for economic gain even if the costs to this are borne by the community at large, including those who have not been integrated into the existing water management systems. The contestations between these stakeholders produce a troubling level of inertia when it comes to initiating actions to resolve problem in the water management system. Some necessary interventions, such as addressing pollution of surface water sources to make them accessible for healthy consumption, have no constituency beyond those forced to rely on the rivers and canals in lieu of connections to sanitary infrastructure. Consequently, those who need it cannot, on their own, afford to have it. Rather than implementing the wastewater management, the government's approach has been to remove settlements along the rivers and canals that lack access to clean water and sanitation. Removal of settlements and dredging might remove some sources of pollution and allow the rivers to handle a greater volume of runoff and rainwater, but it does offer little hope of curtailing flooding as evidenced by the recent experience of January 2020 flood. This strategy, which does not include removing the sources of stormwater intrusion not directly connected to these settlements, cannot address pollution, restore the rivers, or reduce reliance on groundwater throughout the urbanized area. A comprehensive river cleanup, coupled with the implementation of citywide wastewater management, is essential to deal with water quality. One major criticism of the JCDS was that if Jakarta allowed its currently polluted river waters to flow into the proposed polder, the result would be a vast body of fetid waters.

Of course, Jakarta does not control all of the river functions. It lacks authority to control the impacts on the rivers that occur beyond its governmental jurisdiction.

As Goh notes, "At the uppermost elevations of the Ciliwung River, there is ongoing contestation between large tea plantations, landscape restoration activities, and new urban development" and affects the ability to control pollution.[40] The inability to prevent dumping upstream is evident in the Citarum River where similar contestations between agricultural producers, industrial operations, local residential communities, and district and regional governments explains why this key source of surface water for Jakarta has become one of the most polluted rivers in the world.

Missing from analyses of how Jakarta can effectively respond to the causes and implications of the flooding is the consideration of ecological solutions. At one point in the campaign of Joko Widodo for the Jakarta governorship in 2012, he set forth the goal of restoring all of the 13 rivers passing through the city to their original character. He pointed to the example of Seoul where the city restored the Cheonggyecheon River by removing an elevated highway and constructing a 3.6-mile linear park to become the "green heart" of the city. Seoul's government justified the project as a flood mitigation initiative even though its central purpose was to bring a clean and green river to the heart of the city.[41] Jokowi's call for restoring Jakarta's rivers to their original condition got lost in the search for quick solutions to the flood crisis once he got in office. It also suffered from the long-standing tradition of elevating growth and development above preserving the environmental qualities within the city's planning and policy agenda. To effectively embrace the strategy of "room for the river" requires protection of the hydrological assets of what once the premier "water city" in Southeast Asia. That involves a concerted effort to restore the rivers. There is in Jakarta, as in Seoul, an opportunity to restore its "green heart" as a potentially more enduring and successful response to the flood crisis than relying on more hard infrastructure as has been the practice. Can it be done? To answer this definitively, first we need to consider how Jakarta got to this juncture, and how might recreate a modern version the water city it started as. But this time, as a more inclusive, more equitable, and more sustainable place. The intent of this study is to offer a historical guide that might lead to that different and desired future for Jakarta.

Organization of this study

This study of Jakarta's battle to mitigate the floods that routinely ravage the city begins with exploring the place where it is situated and the processes undertaken to create a city settlement in a floodplain. Examining the early eras in the city's development sheds light on how and why intense flooding has become such a signature feature of life in Jakarta. In contrast to the seemingly inexorable problem of flooding in present-day Jakarta, the historical record indicates that this was not always the case. Floods of any appreciable amounts were rare in the first several centuries of the city's development largely because the surface waters that increased during the rainy season supported the life of the community rather than destroying it. The community-building process worked in harmony with the natural processes, aided by engineering that guided the abundant waters where they would best serve it.

The ecological context of the place where the Dutch planted the Batavia settlement influenced the development of Jakarta over its first three centuries. Its natural features, the rich water resources available from the rivers, the proximity of the sea, and the availability of arable lands within the region to support agriculture made this an appropriate place to establish a port by the Dutch in the 17th century. The rivers and the topography of land between the rivers initially defined the spatial limits of urban expansion. Over the past century, however, the rivers and canals ceased to guide the city's development or to serve as a critical water resource. Jakarta's rivers obstructed the city-building process and except during floods became invisible within the landscape. Only when the annual rains brought waves of water from the highlands south of Jakarta into the city center did these slow-moving fetid streams devoid of any beauty take on their historic form. During those annual episodes, as they sought their original path through the landscape, they conflicted with the urban structures that had invaded their space. Rather than assets, they were now agents of destruction, powerful enough to shut down the modern megacity. To understand the important relationship between water and the city, Chapter 1 offers a way to see the relationship between Jakarta's 13 rivers and the patterns of growth in the city. It will help to make clear how they changed as the built environment of the city changed.

Given the hydrological assets of the site, it is important to know how the original settlers planned and built the "water city" of 17th and 18th centuries and how Batavia evolved into the megacity of Jakarta of the 21st century. The founding of Batavia by the Dutch on the conquered remains of the Sundanese settlement of Jayakarta required harnessing the river system just like in Holland where canals served as a lifeline for its main cities. Their efforts to fashion a prosperous tropical city proved successful, prompting visitors to crown Batavia the "Queen City of the East." Dutch water engineering played the determining role in shaping its urban form and ensured the successful functioning of an administrative and trade center planted in such a watery environment. Batavia's administration carefully managed the hydrological features of the Batavia and its hinterlands and proved able to keep the inhabitants of this water city "above" water for much of the first three centuries of the city. Four concerns dominated the Dutch approach to water management. These included connecting the river system to a system of engineered canals to support their commercial needs, mitigating the effects of annual flooding through regularly dredging and river management, ensuring an adequate supply of fresh water, and addressing water pollution and its health consequences. Flooding was commonplace throughout the first three centuries of the city. It was, however, of much less consequence since keeping the canals dredged to support commerce and ensuring enough water for the domestic needs of the growing city population meant that more water was better, and the rivers and canals could handle even the heaviest rainfall on a regular basis. There were notable changes in the community's relationship to its environment beginning in the 19th century as the city expanded inland to higher, drier ground and depended less of the engineered waterways. By the early 20th century, the center of Batavia no longer remained on its original waterfront location, reflecting important changes in the city's relationship to the watery environment. There was far less attention to river and canal management since the major commercial port had shifted to a new site east of the city in the late 19th century. Some new infrastructure afforded flood protection for the European settlements in the 1920s along with improved provision of clean water for the growing population that the traditional river sources could not provide. Dutch management of this city ended with the Japanese occupation of Batavia in 1942 and the subsequent independence of Indonesia in 1950, and with these changes, so went the expertise to manage its waterways. Chapter 2 examines this extended and critical foreground to the emergence of the modern water management and flood control systems in Jakarta.

The rapid growth of the Indonesian capital city after independence in 1950 quickly outstripped the infrastructure facilities installed during the colonial regime to manage the waterways. Waterways previously used as sources of potable water, for small-scale commerce and for recreation, ceased to play those roles except to the extent that they provided services to the informal settlements that new migrants constructed along their shores. These settlements occupied green spaces where the rivers traditionally migrated during heavy rains. Their location and expansion was outside the formal processes of planning for urban growth and never appeared as an issue directed by policy. Not until completion of Jakarta comprehensive plan in 1965 was there any mention of environmental objectives within the context of land development controls. Flood control was not a serious policy consideration even though it occurred regularly. As in the past, the annual flooding usually occurred at a limited scale and largely affected the informal communities not officially recognized in the plans. Chapter 3 discusses the water and flood risk management conditions as planning for Jakarta began to consider the implications of the emerging metropolitan complex in the 1970s. This larger region, commonly referred to as JABOTABEK (derived from the first letters of Jakarta and the surrounding cities of Bogor, Tangerang, and Bekasi), became the new focus of water management planning. Preserving open space to support groundwater recharge in the face of rapid urban expansion was one key environmental objective of the Jabotabek planning, an objective consistently ignored in practice. Jabotabek plans acknowledged that effective river management required regional interventions, and that the flooding problem needed a regional approach. The recommended improvements presented in the donor funded studies were routinely neglected by the authorities because they required major investment but also because there were more pressing priorities than preventing flooding that typically hit hardest in the poor communities. Neglected improvement identified in the donor-funded studies included wastewater treatment, stormwater management, flood control infrastructure (including completion of the original Dutch flood canal system), and expanding the availability of piped potable surface waters to the city's rapidly growing population. Several studies noted that groundwater extraction contributed to land subsidence, but this finding failed to be linked by the analysts to the problem of flooding. Recommendations to regulate the location of new development to protect water recharge areas was among the most important suggested policies, a recommendation routinely ignored in practice in favor of continued growth. Although water pollution was a concern, and justified calls for expanded wastewater treatment, no serious plans emerged to restore and to protect the quality of the river water that coursed through the metropolitan area and was steadily deteriorating in quality.

Concerns over unregulated urbanization and overheated development in the peripheral areas of Jakarta produced a pivoting of the development community toward the long-neglected historic waterfront areas. Leading Jakarta developers recognized its potential to complement the profitable projects already underway in the periphery. The rediscovery of waterfront Jakarta and its potential to continue to fuel urban development began in the 1980s as several Jakarta developers recognized, as did their counterparts in other Asian cities, the potential of high-end development in waterfront settings. A comprehensive waterfront development initiative, sanctioned by the government under the Suharto New Order regime, transformed a collection of private real estate projects into a national effort to create a world class waterfront city on the foundations of the city 17th-century commercial port. The Jakarta Waterfront City project surfaced during the height of the city's development frenzy in the early 1990s. Although the real estate bubble burst in 1997, contributing to a decade-long economic collapse throughout Indonesia and the subsequent demise of the powerful Suharto government, the waterfront redevelopment scheme remained alive as a vision for Jakarta to realize the dream of gaining world city status. Concurrent with this final decade of the New Order government was the onset of a period of suddenly devastating floods. One reason the waterfront city idea stayed alive is because some recognized it as a flood mitigation and environmental upgrading strategy. In its initial conception, the Waterfront City project envisioned a new environmentally superior "heart for Jakarta" with comprehensive clean water and wastewater service accessible to the majority of its residents and sparkling rivers flowing into the expanded and modernized waterfront. Chapter 4 examines the coastal development initiative of the 1990s to reconnect the city to its waterfront heritage. The plan was to be a catalyst to environmental upgrading through the coastal floodplain since virtually all of Jakarta's rivers would flow in this new city. This included discussions of river cleanup and a shift to privatization as an answer to Jakarta's inadequate clean water service. Underlying the waterfront scheme and water privatization was the assumption that these would reduce the city's dependence on groundwater.

The historic floods beginning in the 1990s that continued through the first two decades of the 2000s, along with the governmental responses and political implications, receive detailed coverage in Chapter 5. Beginning with the governorship of Sutiyoso in the mid-1990s, support for cleaning up Jakarta's riverfronts, eliminating illegal inhabitation along both the rivers and canals, and implementing new strategies to counteract the annual flooding during the rainy season gained growing political urgency. Recognizing that flooding was more than just the result of too much rain, but also resulted from the loss of river capacity to handle the annual rainfall and runoff, became part of the post-flood assessment. The impact of land subsidence owing to excessive extraction from the city's groundwater sources came to light in the aftermath of the 2007 flood. Despite unassailable evidence that the overreliance on the groundwater was the fundamental problem, initiatives to shift water consumption to surface sources failed to materialize. To improve river flow capacities and to eliminate what government regarded as a key source of water pollution, a process of clearing out informal waterfront settlements along Jakarta's river and along the north coast took place in conjunction with a comprehensive river and canal dredging project. Within the context of newly democratized politics in Indonesia, the informal communities affected by the dredging efforts resisted displacement and with support from an array of community organizations challenged the assumption that they were the leading cause of the flooding and the polluted waters.

The massive and long overdue river dredging effort began just as the government released a bold new version of the waterfront development scheme, known as the Jakarta Coastal Defense Strategy. Through a process of land reclamation along the Jakarta north coast (as originally proposed in the 1990s), constructing expansive polders (holding ponds) to accommodate floodwaters, and restoring the natural flow of the city's major rivers, Jakarta would manage its floodwaters much as it was so successfully done in the Netherlands. The plan was unveiled in 2012 at the start of the governorship of Joko Widodo. At the same time, in conjunction with the Widodo administration's intention to address environmental problems, Jakarta launched a series of greening projects and transportation upgrades to support these new land development practices. The JCDS was not just a flood control effort but in fact a reformulation of the 1990s waterfront city comprehensive urban development strategy. The scheme assumed a commitment to removing the seemingly inexorable flow of debris in Jakarta's river system. By cleaning the rivers, reducing reliance on subsurface water sources, and constructing the new city on land not affected by subsidence, the JCDS promised not less than realization of a brand new, problem-free, environmentally upgraded water city. Almost immediately following the public release of the plan in 2012, the JCDS faced the same intense backlash from environmental and community groups that greeted the initial waterfront city project in the mid-1990s. While the land reclamation component of the scheme was the focus of opposition, its links to river dredging to achieve river improvements (or river "normalization") and to address flooding along the north coast gave it broader political appeal that the original concept lacked.

In effect, river "normalization" was not river restoration in the mode of a greening strategy but rather a combination of dredging, rebuilding or constructing channel walls, and removing the illegal settlements that occupied these spaces. Chapter 6 examines this process from the standpoint of the communities affected by this flood mitigation effort and those who bore the real costs of this flood mitigation strategy. As previously noted, these communities were key stakeholders in the political contestations that accompanied efforts to address flooding in the sinking city of Jakarta. There were cases of riverfront communities successfully negotiating at least a temporary stay from removal, but there were many more cases where the government systematically removed communities without compensation appropriate to the losses endured. Evidence also suggests that the policy objectives of flood mitigation were mixed with equally powerful imperatives to remove vestiges of Jakarta's long-standing status as a city of poor villages, an image that undermined efforts to showcase it as the global metropolis it desired to become. Perhaps it was the realization that this goal was too difficult to achieve given the current array of environmental deficits that explains why the government decided to shift its focus to a new "green" capital city in Borneo.

The piecemeal and incomplete interventions to develop an effective flood risk management system after more than three decades of trying suggests that different approaches may be necessary to achieve a more environmentally sustainable megacity. Concluding Chapter 7 examines alternatives to address land subsidence, the impact of climate change, the continued problem of the polluted rivers, the social dislocations associated with river normalization on the city's most vulnerable residents, and more effective ways to manage the flood problem through restoration of a balance between the city and its environment. It explores alternative water management strategies in light of the enduring legacy and unique environment of this water city.

A final objective of this study is to assess Jakarta's efforts to develop effective flood management in the context of recent scholarship that offers critical perspectives on flood risk management drawing upon global experiences. There has been a growing body of studies since the era of regular mass flooding in Jakarta began after 2002, each examining different dimensions of the Jakarta flood problem and the responses. Several studies forge connections between the colonial era focus on engineering responses to flooding and the current efforts to embrace similar infrastructure improvements as the best answer to the problem.[42] Another new line of scholarship points to shifting settlement patterns during the colonial era that affected Jakarta's capacity to provide residents with clean water from surface sources. This led to the unsustainable reliance on groundwater.[43] Demonstrating the relationship between land subsidence, excessive groundwater extraction, runaway urban development, and increased flooding is a central finding of research that points to the sinking of Jakarta and other cities throughout the region[44] The excessive reliance on groundwater is discussed in terms of the inability to utilize polluted surface water sources[45] and also how otherwise accessible sources of the clean water distributed by the water companies do not reach a large segment of Jakarta's population. This is a result of its cost and the lack of connections to water systems.[46] Other analyses examine tools used to anticipate the impact of climate change and to assess proposed flood control interventions, especially the utility of a seawall.[47] Research studies that examine the policy responses to flooding, the problem of surface water pollution, and the impact on the poor communities that reside along these flood-prone polluted waterways underscore the social and economic costs of interventions.[48]

Why examine Jakarta's water management history in such detail as done in this study? The most obvious reason is that water and the management of water has been imbedded in the form and function of Jakarta since its inception. Jakarta has a much different relationship to water than other coastal cities, such as New Orleans, London, Ho Chi Minh, Manila, or Shanghai, all of which were founded on large navigable rivers that were easily harnessed to support the city's development. In Jakarta's initial settlement as the colonial enclave of Batavia, none of the 13 rivers alone were substantial enough to sustain a city. Indeed, even the smaller ships of the 17th and 18th centuries had to anchor offshore. So harnessing these small rivers into an integrated collective water system through engineered infrastructure was what made Batavia work as a city that served as the administrative center for the vast Dutch East Indies Empire. Because of the predictability and reliability of the annual rainy season, the river system continuously nourished the settlement and, for two centuries, delivered all the water that was needed. Protecting and managing those waters to support the urban settlement was a top priority. When Jakarta ceased to need its waterways and stopped managing them as had been done previously to guide urban expansion, was when they ceased to function as the city's vital asset. Throughout the 19th and most of the 20th centuries, waterfront Batavia/ Jakarta assumed the status of the city's backwater. Attempts to restore the glory of the water city in the late 20th and early 21st centuries reestablished a connection to that neglected past and initiated an exploration of how to situate a modern megacity in a floodplain and make positive connections to the environment. An examination of the city's historic connections to water may help to identify strategies to meet the challenges of the present and shape a future in ways that serve its millions of citizens. That is the intention of the story that follows.

Notes

1 *Jakarta Post*, January 29, 1990, p. 2.
2 Shatkin, Gavin (2019) "Futures of Crisis, Futures of Urban Political Theory: Flooding in Asian Coastal Megacities," ***International Journal of Urban and Regional Research***, 43 (2): 217. doi: 10.1111/1468-2427.12758; Octavanti, Thanti and Charles, Katrina (2019) "Evolution of Jakarta's Flood Policy Over the Past 400 Years: The Lock-in of Infrastructural Solutions," ***Environment and Planning C: Politics and Space***, 37 (6): 1102–1125; *Jakarta Post*, January 1, 2020.
3 Darmono, Nglinting (2002) "Indonesia: Floods Exacerbate Economic Disaster." www.greenleft.org.au/2002/482/28712.
4 Caljouw, Mark, Nas, Peter J.M., and Pratiwo (2004) "Flooding in Jakarta," paper presented at First International Urban Conference, August 23–25, Surabaya, pp. 6–13.
5 Ibid., p. 9.
6 ADPC (2007) "Environmental Recklessness Blamed for Jakarta Floods." www.ens-newswire.com/ens/feb2007/2007-02-12-01.asp.
7 Tambunan, Mangapul P. (2007) "Flooding Area in the Jakarta Province on February 2 to 4, 2007," Paper presented at 28th Asian Conference on Remote Sensing, Kuala Lumpur, November 12–16, p. 6. Tambunan borrows data directly from the Caljouw, Nas, and Pratiwo article about the 2002 flood but incorrectly attributes these figures to the 2007 flood.

8 Arambepola, N.M.S.I. and Iglesias, Gabrielle Rosales (2008), "Effective Strategies for Urban Flood Risk Management," paper presented at the Economic and Social Commission for Asia and the Pacific and the Pacific Expert Group Meeting on Innovative Strategies Toward Flood Resilient Cities in Asia-Pacific, p. 1. http://researchgate.net/publication/324560981-Effective_strategies_for_urban_flood_risk_management.
9 Davies, Richard (2015) "Jakarta Floods Force 6000 Evacuations," *Asia News*, February 10. Accessed December 17, 2019: https://floodlist.com/asia/jakarta-floods-force-6000-evacutations.
10 "Torrential Rains Flood Indonesia," *Earth Observatory, NASA*, January 2, 2020. http://earthobservatory.nasa.gov/images/146113/torrential-rains-flood-indonesia.
11 World Bank (2011) *Jakarta Case Study Overview: Climate Change, Disaster Risk and the Urban Poor: Building Resilience for a Changing World*. World Bank: Jakarta.
12 Pelling, Mark, Blackburn, Sophie, et al. 2013. "Case Studies: Governing Social and Environmental Transformation in Coastal Megacities," in Pelling and Blackburn, eds. *Megacities and the Coast: Risk, Resilience and Transformation*. London: Routledge, p. 203.
13 Douben, N. and Ratnayake, R.M.W. (2013) "Characteristic Data on River Floods and Flooding; Facts and Figures," in Alphen, Jos van, Beek, Eelco van, and Taal, Marco, eds. (2006) *Floods, From Defense to Management: Symposium Proceedings*. London: Taylor & Francis, pp. 20–21.
14 Brinkman, Jan Joop and Hartman, Marco, cited in Shatkin, op. cit., p. 218.
15 Yeung, Yue-man (2010) "Coastal Mega-Cities in Asia: Transformation, Sustainability and Management," *Ocean and Coastal Management*, 44 (5–6): 319–333; Blackburn, Sophie and Marques, Cesar, "Mega-urbanisation on the coast: Global context and key trends in the twenty-first century," in Pelling and Blackburn, op. cit. Table 1.1, p. 3; Angel, Shlomo, Parent, Jason, Civco, Daniel L., and Blei, Alejandro M. (2012) *Atlas of Urban Expansion*. Cambridge, MA: Lincoln Institute of Land Policy; Daigle, Katy (2015) "Asia's Coastal Cities Face Challenge of Rising Seas," December 10. http://phys.org/news/2015-12-asia-coastal-cities-seas.html.
16 Storch, Harry and Downes, Nigel K. (2011) "A Scenario-based approach to assess Ho Chi Minh City's urban development strategies against the impact of climate change," *Cities*, 28 (6): 517–526.
17 Nguyen, Qui T. (2016) "The Main Causes of Land Subsidence in Ho Chi Minh City," *Proceedia Engineering*, 142: 336–340.
18 World Bank (2018) Flood Risk Management in Dhaka: A Case for Eco-Engineering Approaches and Institutional Reform. Washington: World Bank.
19 Ibid.
20 Sengupta, Somini and Manik, Julfikar Ali (2020) "A Quarter of Bangladesh is Flooded: Millions Have Lost Everything," *New York Times*, July 30. Accessed December 16, 2020.
21 Fuchs, Roland J. (2010) "Cities at Risk: Asia's Coastal Cities in an Age of Climate Change," *AsiaPacific Issues*, 96 (July): 1–11.
22 Blackburn and Marques, op. cit., pp. 3–4.
23 Abidin, Hasanuddin Z, Andreas, Heri, Gumilar, Irwan, Fukuda, Yoichi, Pohan, Yusuf E., and Deguchi, T. (2011) "Land Subsidence of Jakarta (Indonesia) and Its Relation With Urban Development," *Natural Hazards*, 59 (3): 1753–1771.
24 Fuchs, p. 2.

25 Phien-wej, N., Giao, P.H. and Nutalaya, P. (2006) "Land Subsidence in Bangkok, Thailand," *Engineering Geology*, 82 (4): 187–201. doi: 10.1016/j.enggeo. 2005. 2005.10.004.

26 Fuchs, pp. 2–3.

27 Firman, Tommy (2000) "Rural to Urban Land Conversion in Indonesia During Boom and Bust Periods," *Land Use Policy*, 17 (1): 13–30; Firman, Tommy and Fahmi, Fikri Zul (2017) "The Privatization of Metropolitan Jakarta's (Jabodetabek) Urban Fringes," *Journal of the American Planning Association*, 83 (1): 73–74.

28 Padawngi, Rita and Douglass, Mike (2015) "Water, Water Everywhere: Toward Participatory Solutions to Chronic Urban Flooding in Jakarta," *Pacific Affairs*, 88 (3): 523–524.

29 www.ens-newswire.com/ens/Feb2007/2007-02-12-01.esp.

30 Goh, Kian (2019) "Urban Waterscapes: The Hydro-Politics of Flooding in a Sinking City," *International Journal of Urban and Regional Research*, 43 (2): 250–251. doi: 10.1111/1468-2427.12756.

31 Emmerik. Tim van (2020) "Research: Ciliwung Among the World's Most Polluted Rivers," *Jakarta Post*, February 21. Accessed December 31, 2020: http://thejakartapost.com/academia/2020/02/21/research-ciliwung-among-the-worlds-most-polluted-rivers.html.

32 Leitner, Helga, Colven, Emma, and Sheppard, Eric (2017) "Ecological Security for Whom? The Politics of Flood Alleviation and Urban Environmental Justice in Jakarta, Indonesia," in Heise, Ursula K., Niemann, Michelle and Christensen, Jon, eds. *The Routledge Companion to the Environmental Humanities.* London: Routledge, pp. 194–205.

33 World Bank (2008) Jakarta Urgent Flood Mitigation Project, P1111034. World Bank: Jakarta.

34 Alphen, Jos van, Beek, Eelco van, and Taal, Marco, eds. (2006) *Floods, From Defense to Management: Symposium Proceedings.* London: Taylor & Francis, pp. 5–7.

35 Ibid. p. 6.

36 Pichel, G. (2006) "Jakarta Floods," in Alphen, van Beek, and Taal, op. cit., p. 320.

37 Structure of Jakarta government discussion.

38 Ghalina, Ghina and Tehusijarana, Karina M. (2009) "Borneo, Sulawesi top Capital Candidates," *Jakarta Post*, May 2.

39 Goh, p. 251.

40 Ibid. p. 254.

41 Wang, Lucy (n.d.) "How Cheonggyecheon River Urban Design Restored the Green Heart of Seoul," *Inhabitat.* Accessed January 1, 2021: http://Inhabitat.com/how-the-cheonggycheon-river-urban-design-restored-the-green-heart-of-seoul/.

42 Sedlar, Frank (2016) "Inundated Infrastructure: Jakarta's Failing Hydraulic Infrastructure," *Michigan Journal of Sustainability* 4 (Summer): 1–11; Octavanti and Charles, op. cit.

43 Argo, Teti A. (1999) *Thirsty Downstream: The Provision of Clean Water in Jakarta.* PhD dissertation, University of British Columbia, Vancouver, Canada; Kooy, Michelle and Bakker, Karen (2014) "(Post) Colonial Pipes: Urban Water Supply in Colonial and Contemporary Jakarta," in Colombijn, Freek and Cote, Joost, eds. *Car, Conduits and Kampongs: The Modernization of the Indonesian City, 1920-1960.* Leiden: Brill, pp. 63–86.

44 Abidin et al. (2011), op. cit.

45 Argo (1999).
46 Argo, Teti A. and Laquian, Aprodicio A. (2011) "The Privatization of Water Services: Effects on the Urban Poor in Jakarta and Manila." http://wilsoncenter.org/sites/default/files/Argo.doc.
47 Budiyono, Yus (2018) "Flood Risk Modelling in Jakarta: Development and Usefulness in a Time of Climate Change." PhD Thesis, Vrije Universiteit, Amsterdam.
48 Caljouw, Mark, Nas, Peter J.M., and Pratiwo (2005) "Flooding in Jakarta: Towards a Blue City with Improved Water Management," *Bijdragen tot de Taal-, Land-en Volkenkunde*, 161 (4): 454–484; Leitner, Colven and Sheppard (2017), op. cit.; Leitner, Helga and Sheppard, Eric (2017) "From Kampungs to Condos? Contested Accumulation Through Displacement in Jakarta," *Environment and Planning A: Economy and Space*, 50 (2): 437–456.

1 Water in the urban landscape

Jakarta's water ecology is composed of a dense network of rivers and their tributaries that flow northward from the highland Puncak region to the south, with most ending by depositing into the Java Sea. The existence of these water resources attracted the Dutch to the site and inclined them to establish the port city of Batavia in the early 17th century. There are 13 rivers (*kali*) and their tributaries that form the delta upon which the future megacity would sit. Three of these traverse over 100 kilometers to water the lowlands where the city was fashioned. Three of the four main rivers among the "Jakarta thirteen" are the Kali Pesanggrahan and Kali Krukut, feeding the western side of the urbanized area, and Kali Sunter running through the eastern side. The fourth is Kali Ciliwung that begins high in the mountains south of Jakarta and meanders lazily through periphery of the megacity and terminates in the center of the Jakarta. In the 17th and 18th centuries, the Ciliwung was much grander in scale and served as the central waterway of the Batavia settlement. A fifth major river, the largest and longest in West Java, Kali Citarum, begins far up in the mountains near the city of Bandung. Although the Citarum does not flow directly through Jakarta, it serves as the major source of water for agricultural and domestic consumption for settlements along the northwestern coast of Java, including Jakarta. It is intensely polluted, but none of the other Jakarta rivers possess either the volume or a higher quality of water to help relieve the burden of the Citarum to supply surface water. Because of the volume of water flow, the Citarum helps to run three hydroelectric plants and replenishes the Jatiluhur Reservoir, the largest freshwater source in West Java, and where Jakarta's water company gets much of its surface water.

The 225-kilometer Citarum River, which begins in Lake Cisanti (south of Bandung) flows from South of Bandung to the Java Sea, passes through the eastern periphery of metropolitan Jakarta. It was long connected to the Tarumanegara Kingdom of West Java that ruled the interior lands prior to and after the Dutch conquest. It is the primary source of water for 25 million residents inhabiting nine regencies and three cities in West Java. It also replenishes the Saguling, Jatiluhur, and Cirata reservoirs as well as a major hydroelectric power plant and even irrigates surrounding rice paddies. With all of these life-supporting functions it is noteworthy that the Citarum ranks among the most polluted rivers in the

DOI: 10.4324/9781003171324-2

world. As noted in a 2014 article in *The Telegraph*, more than 200 textile factories line its banks and have "choked the Citarum with both human and industrial waste," including lead, arsenic, mercury, and tons of plastic mingled with bodies of dead fish.[1]

A documentary film produced by two French environmental activists shot as they canoed through the mounds of debris in August 2017 drew a fast and positive response from the Indonesian government. Beginning in November, cleanup of Lake Cisanti started. In February 2018 on orders from President Widodo who had seen the video, the Indonesian military descended with 5,000 troops to begin an extraction process of the mounds of debris in the Citarum. In addition to using heavy equipment to load up the solid waste, the cleanup crew also planted new trees on the riverbanks to slow down soil erosion. Widodo proposed legislation to make it possible to prosecute the major offenders, such as polluting factories and public littering. Troops remained to patrol stretches of the river to support the nonpolluting regulations. Even with the cleanup and some continued monitoring, households with faulty septic tanks that line the river, along with the uncooperative industrial plants, continued to pose a threat to this important river. The major problem is that only a small percentage of the factories along the river connect to the one wastewater management plant actively operating, so the toxic material must go into the river untreated.[2] Along the vast network of tributaries that feed into the Citarum are villages that contribute their waste to what the *Jakarta Post* in January 2019 referred to as the "River of Filth." The Citarum Harum program launched by President Widodo in March 2018 to clean the river will take up to seven years to complete.[3] It is the polluted condition of the water that the Citarum brings to the Jakarta water company that is an obstacle to reducing Jakarta's dependence on groundwater. It is overdependence on groundwater that threatens the future viability of Jakarta.

As the Citarum case demonstrates, it is not just what happens to the rivers as they pass through Jakarta, but also what happens upstream, that influences water quality.[4] This has been the case since the earliest settlements along the north Java coast. Before urbanization spread deep into the interior, this expansive river network passing through the flat lands surrounding Batavia enabled the agricultural enterprises to thrive. Figure 1.1 shows the major rivers that pass through the area that became the city of the 20th century. As urbanization displaced agricultural cultivation on these lands, Jakarta turned to an alternative source of water, namely groundwater drawn from the deep aquifer that sits below the urbanized area. The Batavia government first tapped groundwater in the 19th century, as will be discussed in Chapter 2, but the rivers continued to provide the main source of water for consumption for the majority of urban residents for several decades after independence in 1950. A third source of water for Jakarta was the Java Sea since for at least the first several centuries of its history, it was the sea that flushed the river waters to keep them relatively clean (at least by standards at the time) for use by the residents.

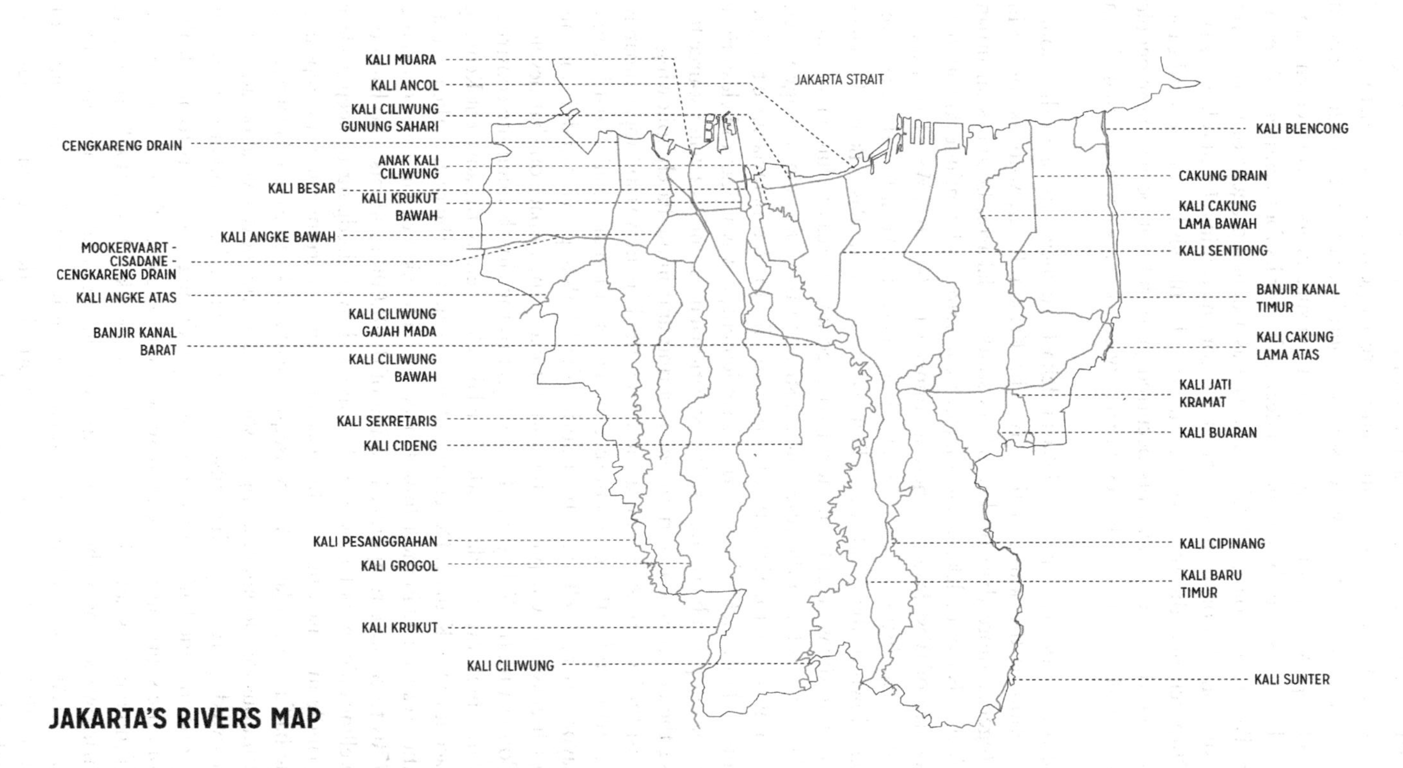

Figure 1.1 Jakarta's main rivers, tributaries, and canals.

Source: Prepared by Elita Nuraeny, Department of Architecture, University of Indonesia.

A fourth source is the rainfall that provides a steady and highly reliable contribution of water to the surface sources and to recharge the aquifers that supplied the groundwater. All of these water sources contributed to the development of the city, all required continuous investment to ensure proper management, and all posed challenges that the city had to deal with over its four centuries of development. The hydroecology of Jakarta served as a defining agent of urban form and the key contributor to the quality of life from its origins. Until relatively late in the 20th century, the rivers of Jakarta were highly visible in the landscape. They influenced where settlements occurred (and did not occur), how people navigated the city, and provided essential service to many seeking a new life in the city.

All of the rivers that form the Jakarta delta have their origins in the mountainous region to the south of the city. The Krukut River, with its source in Bogor Regency more than 60 kilometers south of Jakarta, becomes an urban river as it travels from a lake in Depok (Situ Citayam) and continues through to the Karet area of West Jakarta. There it connects to the city's West Flood Canal in Kelurahan (subdistrict) Kebon Melati. From there, the West Flood Canal takes the waters down to the floodgate at Pluit Dam (Pintu Air Pasar Ikan). It makes one last turn to the west and continues to the reclaimed area of Pluit on the north coast of Jakarta and into the reservoir constructed there to help manage water. The Krukut is one of the rivers that has periodically flooded Batavia since the late 19th century. For much of the life of the settlement, it was a clear and clean river used for consumption by residents, for recreational uses, and as a water source for irrigating the adjacent agricultural fields during and immediately following the Dutch East Indies regime. As late as the 1970s, the Krukut was still approximately 25 meters wide and with a depth of several meters. It ran alongside what became the Kebayoran Baru satellite city built during the 1950s to provide middle-income housing (discussed in Chapter 3) but did not exert any impact on that development. Connecting the Kebayoran Baru project to the city through the construction of a new highway stimulated additional development in the adjacent Kemang area adjacent to the Krukut. Development encroached on the river as the population in the area swelled in the 1960s. Governor Ali Sadikin began a process in the early 1970s to remove temporary housing of the 8,000 residents who had built their dwellings along its banks. He also initiated river dredging to keep it clear and to handle increased flows. This was a one off occurrence and not repeated until one-half a century later. As a consequence of a lack of sustained maintenance and continuous development along its shores, over the next several decades, the river narrowed appreciably. This caused some sections of the Krukut to shrink in width to one and one-half meters, substantially reducing its capacity to handle the flows from the south during the rainy season. One section of the river retained a width of three meters, and this earned the name Krukut Lama (Old Krukut). It was the narrow sections of the Krukut upstream in South Jakarta that surged over their banks in August 2016 and created flash floods in the affluent community of Kemang. In this case, it was not the poor informal settlements but the high-end housing that had been constructed illegally along its shores. Following the flood of 2016, then Jakarta Governor (Ahok) Purnama announced that he would tear down all of the structures in the 3.5-kilometer section of the river adjacent to Kemang where the worst flooding occurred. This meant not only any informal structures located there but also luxury units built without proper authorization. Throughout Akok's brief tenure as governor (from 2014 to 2017), he took an aggressive stance toward "normalizing" the relationship between the built and the river environments that meant removing unauthorized settlements. In the affluent South Jakarta neighborhood of Kemang, however, where some its finer houses encroached on the river like the informal settlements, the city never acted upon the threat.

Following the Krukut from its source on a Jakarta atlas shows clearly how both the formal and informal urban development pinch this waterway along its pathway through the city. Once it passes through the channelized portions of Kemang (where it is the luxury house and apartment construction that contributed to the river narrowing to just three meters), it heads north past Kampung Bangka which straddles its western edge and then onward to the residential areas of Kuningan Barat (West Kuningan). It then goes underground, passing under the city's major business street, Jalan Sudirman, until it connects to the West Flood Canal (discussed in Chapter 3). Although the Krukut disappears from the maps at that point, it reappears again as it branches off from the Kali Cideng just before reaching the Kali Besar (Big River) in North Jakarta. According to the highly authoritative Jakarta Street Atlas, the Krukut again connects with Kali Besar at its northern edge before passing by the Tongkol community (which you will meet later on) and then heads toward the sea.[5] The greatly reduced channel of the historic Krukut belies its potential flood impacts during the rainy season. For the thousands of inhabitants living along its shores, from South Jakarta through Central Jakarta and onward to the North Jakarta, the once green and expansive Krukut is now a shallow muddy stream that brings raging flood conditions during the rainy season.

The Kali Sentiong, which is a tributary of Jakarta's great river, the Ciliwung, runs through communities on the eastern side of the city toward Sunter and has the reputation of being one of the most polluted of the city's thoroughly polluted rivers. It is nicknamed the black river because its inky black waters are the result of absorbing more than its fair share of the 2.5 million cubic meters of wastewater that the city's rivers absorb every day. A 2009 law prohibits companies to dump waste in the city's waterways. The need for a law that is not regularly enforced helps to explain why 96% of the river water in Jakarta is "severely polluted." When Jakarta hosted the Asian Games in 2018, the city administration covered a portion of the Sentiong with a dense nylon net in an effort to reduce the foul aromas emanating from the river in the nearly housing for the athletes.[6] As a result of the uniformly polluted qualities of Jakarta's rivers, it is estimated that only 2% of the city's raw water processed by its water company comes from any of its surface waterways. Jakarta boasts only one facility where wastewater is treated. The Setiabudi Reservoir in Central Jakarta, situated between the intensive business districts on Jalan Rasuna Said and Jalan

Sudirman, is the one wastewater treatment facility which handles just 2% of the wastewater needing treatment. There is a plan by the government for a more comprehensive Jakarta Sewerage System (JSS) that divides Jakarta into 15 zones, each with its own wastewater management facility. The cost and complexity of placing wastewater pipes up to 30 meters below the surface and connecting them to existing settlements has kept the JSS largely in the planning stages, with a commitment (not yet confirmed) to have a couple of zones constructed by 2026. A more immediate, and potentially more viable, response is to make use of small-scale interventions. A model of this took place in the Malakasari neighborhood in Duren Sawit, a community in East Jakarta, where the city constructed a communal wastewater treatment facility in 1998. The facility is managed by the water company.[7] Unless improved wastewater treatment occurs throughout the city, and there are proper places to deposit waste, Jakarta's rivers, like the Sentiong, will continue to run black.

The Grogol River (Kali Grogol) is another of the rivers that passes through what has become some of the densest residential and commercial developments in Jakarta. Situated to the west of the Krukut, and extending 23 kilometers from south to north, like the Krukut it originates in a lake in the suburban city of Depok and heads northward through the luxurious Pondok Indah community (and directly through its famed golf course) located in South Jakarta. From there it continues north until it connects with Kali Pesanggrahan on the western side of the city after passing through another high-end community, the Palmerah Urban Village. It was there in September 2014 that the city tore down 113 illegal buildings, the remains of which added further debris to the already clogged channel.[8] The pollution levels of the Grogol are too high for it to be used as a source of raw water, but it does support irrigation needs of the Pondok Indah Golf course as well as the Senayan Golf course located further north in the heart of the city.[9] Even though the Jakarta government included the Grogol in the mass river dredging project that took place between April and May 2013, it still flooded in 2014. One explanation was that the Jakarta Cleanliness Department, charged with cleaning up the debris from torn down illegal buildings along its course, never fully finished the job. The Grogol runs along the western border of Kebayoran Baru. Because that satellite city sits on higher ground, the Grogol never impacted it the way it does others situated further downstream.

The Cisadane River, another river not officially part of the "Jakarta thirteen," is a 138-kilometer watercourse that begins in Mount Mandalawangi south of Bogor and then heads northward to the satellite suburbs of Tangerang and empties into the Java Sea. The Cisadane and its tributaries support the historic kingdom of Banten, the once famous port city that originally controlled the lands on the northwest Java coast that became Batavia/Jakarta. This historic river traditionally marked the recognized border separating the Dutch East India Company (VOC) and the Banten kingdom. Tangerang is one of five fast-growing satellite cities where much of the suburban development in Jakarta occurred since the 1980s and now is an urbanized area dealing with its own problem of flooding.[10]

Another Tangerang river that historically influenced the development of Jakarta (and is one of the Jakarta thirteen) is Kali Angke. It begins approximately 91 kilometers south in the highlands south of Bogor. Likely named after the Banten Sultan, Tubagus Angke (who ruled in the 16th century), it was at the confluence of the Angke and the Cisadane that the Dutch built a fort in 1657 to secure control over this valuable water source. Between 1678 and 1689, they constructed the Mookervaart Canal to divert water to Batavia to fill its canals. When the problem of insufficient water became a crisis for the settlement in the early 18th century, largely because the agricultural sector consumed so much of the available supply, then Governor-General Durven had the 13-kilometer canal deepened to expand the water quantity, a project that proved less than successful because it also decreased the flow.[11]

The Angke connects in West Jakarta to the West Flood Canal, but has a channel (Saluran Angke) that diverts water that eventually passes under the airport toll road where it combines with the flood canal, passing through Pluit and into the Java Sea.

The 37-kilometer Sunter River flows through the eastern side of the Jakarta delta, meandering through what has become the most densely settled areas of the city. As late as the early 1960s, however, it passed through sparsely populated agricultural lands. The Sunter figured prominently in a plan for Pulo Mas, a new mass housing settlement built in the early 1970s. The plan called for the river to absorb stormwater and the treated wastewater and to carry it to the sea. The project was built but with a much lower density and with individual septic tanks rather than the proposed central wastewater system. What caused the declining quality of the Sunter was a combination of untreated waste from new industrial areas that filled it with untreated waste and runoff from massive new developments, such as the Kalapa Gading community in northeast Jakarta, as well as substantial new residential development in the southeast sections of the river. When Jakarta finally got around to dredging its rivers systematically beginning in 2013, a large portion of the Sunter was included in the "normalization" scheme to remove sedimentation and to construct improved embankments. A positive improvement for those residing along the Sunter was the East Flood Canal finally completed in 2012. What had been one of Jakarta's rivers that consistently experienced flooding, since 2012, the East Flood Canal handled water diverted from the Sunter to reduce the threat from high waters in the neighborhoods it passes through. Like all of Jakarta's rivers, however, its waters register high levels of pollution as it continues to function largely as a waterway useful for handling waste. As a worker from the East Jakarta water body management team reported in May 2019, he watched the water turning blue with industrial waste "flowing into the Sunter River from a community drainage system."[12] It was hard to tell whether it came from one of the houses along the stretch of the river at Cipinang Muara or a nearby factory. But as long as the Sunter and all other Jakarta rivers by necessity take the stormwater as well as the wastewater from homes and businesses, there will be no reduction in the levels of pollution. It is likely that the decision in the 1870s to place the new port facilities in Tanjung Priok where the Sunter reaches the sea influenced subsequent land development patterns up and down the river and sealed its fate as a waste site for East Jakarta's communities.

All of Jakarta's rivers ultimate head to Jakarta Bay and that influences the quality of its shoreline and the water body itself. This became a critical consideration when proposals to redevelop the waterfront began in the 1990s and raised the prospects of the polluted rivers affecting anticipated high-end development, as will be discussed in Chapters 4 and 5. Already there were data available on potential impacts of development related to the geology and hydrology of the basin. A scientific assessment – the Verstappen study – of how the Jakarta Bay and the adjacent shoreline developed in the pre-settlement period was conducted by Dutch experts examining data from the 1870s through the 1940s. It verified that the original Batavia settlement had been situated at the lowest point of a break in a beach ridge extending in an arc from east to west, a ridge that had been formed by volcanic eruptions. The Verstappen study contended that the major rivers that contributed to accretion of new lands along the shoreline were largely to the east and the west of the site of the original Betawi settlement. Owing to the relatively modest size of the main waterways passing through what became the center of the urbanized area, the process of carrying down silt to construct new land areas was modest in comparison to those parts of the shoreline served by the larger Citarum and Cisadane rivers. In other words, the lands to the east and west would be less vulnerable to rising waters than the site selected for the original settlement. What Verstappen's study also graphically illustrated is that there had been a substantial drying up of the rivers emptying into the coastal area. This created the marshy lowland areas where much of the later settlement occurred. The alluvial plain that extends from Bogor to the coast area sustained its two major rivers, the Angke and the Ciliwung, on a fixed course, thereby providing a logical north to south corridor for expansion of the settlement and, in fact, that is where Batavia developed from the 17th through the mid-20th centuries. But as the Verstappen study points out, the land upon which Batavia was settled initially previously sat under water. The area in 17th-century settlement was just one and one-half meters above sea level on land where the sea level previously was between four and six meters higher. The dropping of the sea level deposited landmaking material on the alluvial plains and in this way contributed to building higher grounds and terraces along the rivers especially further south toward Bogor. Yet, even so, the land upon which Batavia was founded left little room for additional water to be accommodated.[13] And in the 20th and 21st centuries, with rising sea levels, the potential for this now intensely urbanized site to return to its earlier "under water" condition was very real.

The Ciliwung (Ci Wung or Tjiliwoeng in Dutch) is the river most directly associated with Batavia from its beginning and which continues, in an altered manner, to transect the city center today. Its source is the 3,000-meter Mount Mandalawangi south of the city of Bogor, a range tall enough that on a clear day it can be seen 60 miles away in Jakarta. The Ciliwung runs approximately 97 kilometers to the Java Sea. It drops sharply after its first 17 kilometers in the highlands, which provides the force to propel its waters to Jakarta. The final 60 kilometers of the river traverse largely ground-level spaces. This lack of pitch accounts for the meandering form as it traverses the hinterlands and continues into the central sections of Jakarta. Historically, it was Jakarta's grandest river and served as the avenue of transport and the main water source in the precolonial and colonial eras for the settlement along the coast and villages that grew up along it.

It was populated as early as the 4th–5th centuries on the upper section by the kingdom of Tarumanegara and the lower harbor area by the Hindo Sunda kingdom of Pakuan Pajajaran. For the Dutch colonial rulers beginning in the 1600s, it was a vital resource. They vigorously protected the upper portion of the Ciliwung through a series of outposts along the adjacent road between Fort Pajajaran (Bogor) and Batavia, because of its value as waters for irrigating crops as well as to support the settlement downstream. Where it reached the outskirts of Batavia, the Dutch diverted it to provide the primary source of water for the city's canal system. The Dutch engineered its natural undulating shape into a system of straight canals when it reached the northern edge of what became the main European center of Batavia in the 19th century, the area known as Weltevreden. Because of its central location within the expanding capital city of Jakarta after 1950 and the flat lands along the lower portions of the river, a substantial portion of the growing city population chose to create their settlements along its shores. Current estimates suggest that as many as four million residents still reside along the Ciliwung beginning on the outskirts of the Bogor subdistrict of Ciawi and extending all the way to Merdeka Square in Central Jakarta. The vast catchment area of the Ciliwung (476 square kilometers) explains why during the rainy season it takes on such a robust character. Upstream the Ciliwung still supports agriculture but also is diverted for drinking water by the water companies in Bogor, Cibinong, Citayam, and Depok (adjacent to Jakarta). Once it reaches Jakarta, it is so polluted that it cannot be used for drinking water. Its main function in the highly urbanized areas is to handle urban runoff and as a disposal site for residents of the kampungs. As a result of the pressures of urbanization reducing the width and depth of the river, the Ciliwung has become a primary contributor to Jakarta's persisting flood problems. Cleaning up and restoring the Ciliwung to its historic grandeur figures prominently in the ongoing efforts by Jakarta to manage its rivers. Jakarta's main flood control administrative unit bears its name, the Ciliwung-Cisadane Flood Control Office. Since the early 1990s, virtually every Jakarta governmental regime included "cleaning up the Ciliwung" in its public pronouncement of environmental priorities. Joko Widodo, Jakarta's governor for two years (until elected to the Indonesian presidency in 2014), advocated for restoring not only the Ciliwung but all 13 Jakarta rivers to their original state. His vice-governor and successor, Ahok, went so far as to suggest the goal of cleaning the Ciliwung to such an extent that residents could safely swim in it. To this end, he created a workforce of 4,046 to clean the rivers, lakes, and coastal areas. This "orange uniformed" army had the task of removing what was typically 400 tons of waste per day from these water bodies.[14] That there was on average 400 tons of debris to remove daily itself suggests that the vision of children safely swimming in the river was a bit of political hyperbole.

The main reason for the continued degradation of Ciliwung is the development that has occurred along its long course through Jakarta's periphery. Between 1970 and 2000, undeveloped land in the Ciliwung catchment area dropped from 66% to 38% of the total land area. Over the next decade, virtually all of the remaining land capable of absorbing rain and runoff was no longer vacant.[15] Owing to its centrality and the challenges it faces because of the intensity of development, none of Jakarta's rivers receive the level of assessment bestowed on the Ciliwung. These studies have focused on strategies to restore it as the asset it was in the past. Because of the millions who live on or near the Ciliwung, however, restoring it to the urban asset it once was could unduly burden those who reside there. The complexities of the restoration saga are a story best saved for fuller discussion later on.[16]

Managing Jakarta's complex system of waterways necessitates an integrated system that balances the natural systems with the human intervention associated with urbanization. The approach to managing Jakarta's water depends on who is making the decisions.[17] Decisions can emanate from the national level because Jakarta is the capital city but also because Indonesian law recognizes the authority of the national government over all of its waterways. It is a national regulation, for example, that stipulates all structures must be situated at least ten meters back from any water body. This regulation is the rule without regard to any peculiar local circumstances that might make it inappropriate or perhaps unenforceable. On the other hand, the Jakarta government bears responsibility for day-to-day maintenance of the water bodies within its jurisdiction even though the national government, in essence, holds title. In collaboration with national agencies, local government enforces regulations on the use of these water bodies (rivers, lakes, canals, and even subsurface water) and can impose its own regulations, such as requiring infiltration wells to help restore groundwater sources or preventing development that encroaches on water sources. Subunits within Jakarta also exercise authority over key components of the water and waste management systems. In addition to the array of governmental bodies that set the rules on water resource utilization, another key group of decision makers is the urban consumers. Jakarta's consumers (both individuals and businesses) chose from various sources of clean water, manage their disposition of wastewater, and influence the functioning of the water management system through daily practices, whether legally or outside the accepted norms of proper use. Even those consumers who practice "outside the accepted norms of proper use" are not operating surreptitiously but with the seeming full consent of government. Throughout the kampungs lining Jakarta's rivers, it is routine to find the wastewater pipes extending from inside the house directly over the waterway. Jakarta's formal communities use ditches in front of their homes and businesses to channel the gray water into the nearby streams since Jakarta has no system of pipes to carry stormwater except into the rivers.

By their very nature, rivers possess self-maintenance properties that generate changes regardless of human interventions or being free of these impacts. The morphology of the rivers changes continuously based upon the amount of rainfall and runoff from adjacent land. The flowing of the water itself can reshape the riverbanks and the vegetation that they support. Much of this change, especially within urban areas, occurs because of what happens upstream since upstream activities influence the volume and velocity of the river's flow and introduce sediments, debris, and other foreign materials that can alter water quality and affect the water's ecology. Over time, rivers also naturally create their own floodplains that can help to preserve natural riparian and aquatic ecosystems but also function as floodways during high-water events to protect development near to, but not encroaching on, the designated floodplain. The dynamics of riparian and aquatic ecosystem vitality are best served when the river is allowed to flow where it wants to flow.

The human populations and development they create utilize the resources of the river and affect both the quality and quantity of its waters. During the colonial era in Jakarta, the built environment went right up to edges of the rivers and canals to serve as the primary transportation arteries to move people and goods throughout the city as well as to serve the water consumption needs of the inhabitants. Lands outside the urbanized area remained sparsely developed and there the rivers ran freely to irrigate the crops as well as a source of clean water for consumption. Until the early 19th century when Batavia's European population moved southward from the original coastal site to higher ground in the area known as Weltevreden, living along the riverbanks was desirable. As the city's population shifted southward throughout the 19th and 20th centuries and no longer relied on the rivers for transportation or for domestic water consumption, the riverfront lands lost their desirability. These lowland green areas witnessed the construction of the informal settlements that relied upon the river, not so much for irrigation of crops as to use the rivers for basic water and sanitary services not provided as they were to formal communities in the city.

These developments reshaped the rivers by several different ways. Encroachment of settlements on the riverbanks reduced the river widths. Erection of barriers to reduce erosion and to channel the flow of water to protect adjacent buildings that also went right up the edge also reduced river widths. Figure 1.2 offers a birds-eye view of the canal system built by the Dutch that impinged on the natural systems over the next four centuries. Built into these engineered barriers along the rivers were drainage pipes that allowed surface runoff to add pollutants to the water. These engineered barriers along the banks also further reduced the flow capacity of the channel while increasing the speed of waters that contributed to flooding during high rain events.

During the first two centuries in Batavia, the settlement regularly dredged the rivers and constructed canals since they were the lifeline for the community, and it was necessary to keep these channels to the sea cleared from sediments that flowed down from upstream. Fast-forwarding to the modern era, the irregular efforts to remove surface debris were not complemented by demands (and hence no efforts) to dredge the sedimentation that naturally accumulated daily from the long downstream trip for the water coupled with erosion on sections of the

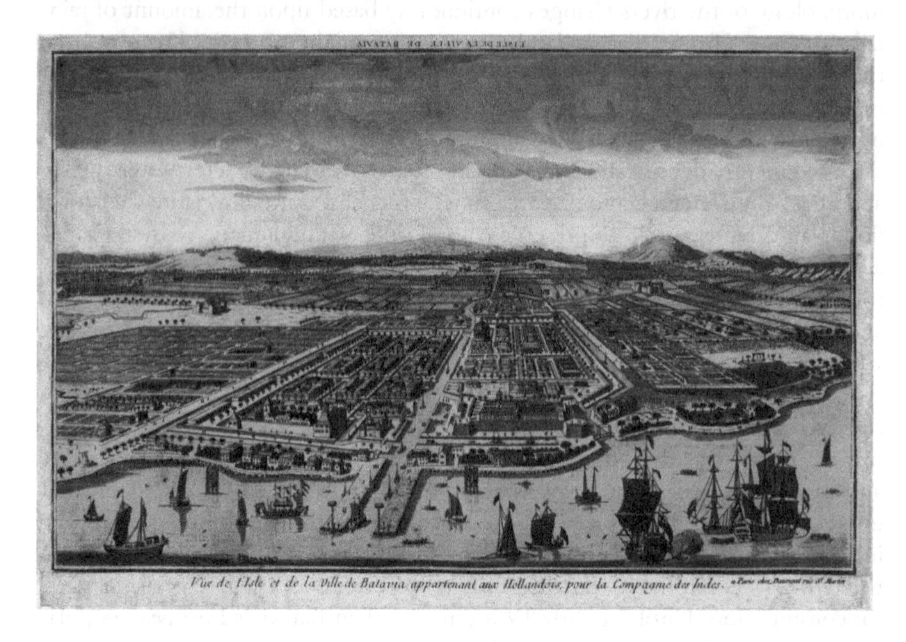

Figure 1.2 Sketch of the settlement of Batavia in 1780.
Source: Collection of the Nationaal Museum van Wereldculturen. Coll. no. TM-3728-537.

rivers where there were no engineered barriers. What were once deep rivers became waterways often just a meter or less in depth, polluted by wastewater from upstream and adjacent settlements and daily inundated with the surface drainage from Jakarta's streets engineered to deposit stormwater in the rivers. By 2020, an analysis verified that all of the rivers passing through Jakarta were thoroughly polluted, in essence open sewers that are treated that way not simply by those who still relied on them for basic needs, but by the city as whole.

Colonial Batavia residents had used the rivers for all their needs, not only for transportation but also as a source of water for consumption and even as a place to deposit their waste. Because of regular maintenance, the rivers and canals were able to self-clean because upstream water, especially during the rainy season, carried the waste to the sea. The dry season was another matter, however. Without the water necessary for the self-cleansing functions to work, this turned the settlement into a smelly and disease-prone place. This encouraged urban settlements to shift inland where the river water was better, but also where clean water drawn from groundwater sources supplanted the river water. The primacy of the rivers and canals to support the city's trade function ceased with the construction of the Tanjung Priok port facility on the shoreline to the east of Batavia. Tanjung Priok could accommodate the large ships that previously needed an anchor off the coast and transship goods to land. As a result, the smaller watercraft that used to carry the good from ships anchored out in Jakarta Bay into the canals and rivers were no longer as important to local commerce as they had been previously. The rivers and canals continued to provide water for the domestic needs of those lacking access to groundwater sources as well as the only citywide wastewater facilities. By the 20th century, all of Jakarta's rivers and the few original canals that had not been filled in to create roads became open sewers, still used by the growing poor population for washing clothes and bathing but utilized by all as the place to deposit the wastewater. In times of high water during the rainy season, they spread their polluted water on the communities that hovered so close to their shores.

Most of Jakarta's polluted rivers feed into the Jakarta Bay. As a result, the area of the original city settlement along the northern coast suffers from some of the most polluted waterways in the city. Here along the shores, the water runs inky black. This urbanized area encompasses a 72-kilometer shoreline along the Jakarta Bay that enjoyed the protection of a dense coastal mangrove swamp when the Dutch planted the city of Batavia there in the early 17th century. The mangroves are gone, cut down to make way for coastal residential and commercial developments. Offshore coral reefs that supported a robust fishing industry have suffered as well. Recent studies verify that erosion of the coral reef over the past 40 years has reduced the fish population. The vast array of urban infrastructure that reduced the mangroves by over 80% left the coral reef vulnerable to decay.[18]

The underground water sources below the delta upon which Jakarta sits provided a replacement for the surface waters that became increasingly unacceptable for domestic consumption beginning in the 1970s. The groundwater head of the deep aquifer below Batavia in 1900 was between 5 and 15 meters above sea level where it begins roughly 60 kilometers to the south of the coastline at the point where there is bedrock under Bogor. One part of the aquifer begins there, separated by a row of sediment barriers (the Tangerang High in the West, the Depok High directly south of Jakarta, and the Rengasdengklok High in the East) where the sediments are up to 300 meters thick. This prevents recharge from getting to the coastal zone. There are five clusters of aquifers situated at different depths below Jakarta, from 0 to 40 meters, 40 to 95 meters, 95 to 140 meters, 140 to 195 meters, and 190 to 250 meters. As Jakarta's water use shifted from predominant reliance on surface water. Subsurface sources experienced heightened demand rapid population and unregulated agricultural and industrial uses, coupled with the loss of permeable surface area traditionally used to recharge the shallower clusters produced a rapid drop of the water supply in the aquifer. This contributed to the land subsidence that became evident by the 1970s. A study of groundwater resources conducted between 1983 and 1985 by a team of German and Indonesian experts found brackish water in the aquifer at least five kilometers inland from the sea because of the drop in aquifer water level.[19] Between 1900 and 1940, the estimated average annual extraction from the aquifer was ten million cubic meters. By 1985, groundwater extractions increased to 47 million cubic meters per year. As noted in the 1985 groundwater study, the horizontal inflow across the ridgeline extending from Tangerang through Depok to the Rengasdengklok High was unable to compensate for greater amounts withdrawn. Over the next several decades, the volume of groundwater extraction increased dramatically to approximately 250 million cubic meters per year, most coming from shallow wells in homes but almost one quarter of it from the 2,000 deep wells.[20] The extraction from the deep aquifers was the key factor behind the drop in the water level of the aquifer and land subsidence throughout central and northern portions of Jakarta to the east and west. In 1985 a depression area of 15 meters was detected that by 2008 had expanded to 25 meters, contributing to a land subsidence rate of between 15 and 25 centimeters per year. Because much of the groundwater was cheap and clean, without regulations there was no incentive to curtail its use.[21]

The excessive reliance on groundwater for domestic, commercial, and industrial consumption, apart from bottled water drawn from highland sources, is because the vast supply of surface waters in Jakarta is highly polluted. Jakarta's canals and rivers, once a source of water as well as the transportation arteries of colonial Batavia, have become the waste receptacles of the modern megacity. At one of the floodgates linking the Ciliwung River to the West Flood Canal, city waste collectors in 2016 extracted "an estimated 740 cubic feet of garbage and natural debris from the river each day," enough to fill three trucks.[22] Bejo Santoso, one of the team of trash collectors at the Manggarai Sluice, traps the floating debris with buoys strung across the waterway. Upstream, another team of workers with heavy equipment removes sediment to enlarge the river channel. Both the trash collectors and the sediment removal processes began in 2012 when Jakarta secured a US$189 million grant from the World Bank to unclog the rivers in an effort to reduce flooding. This represented the first consistent effort since the 1970s to unclog Jakarta's rivers and canals. In this upstream section of the Ciliwung, the dredging removed approximately 35,000 cubic meters of sediment each day to achieve the objective of restoring the river to a depth of 20 feet (or six meters).[23] The story of this renewed dredging effort deserves (and will receive in Chapters 5 and 6) a fuller treatment since its implications extend well beyond the seemingly beneficial process of cleaning up the rivers. River normalization has been the catalyst to a social and physical transformation in Jakarta kampung communities that has pitted government and civil society in a battle over who resides in the city. Critics of the forced eviction of Jakarta kampung residents situated along the rivers and canals being dredged contend that it is not just environmental objectives or flood mitigation that are the reasons for river improvements. It is nothing short of struggle over who has a right to the city. The once accepted practice of enabling migrants to Jakarta to create their own communities on marginal lands is no longer acceptable. How Jakarta redevelops this cleared waterfront areas as green spaces, as new affordable housing, or for other urban uses is a central issue in the political struggle pitting the affected communities against the government.[24]

At the heart of the problem of identifying the future role of Jakarta's rivers in what is an ever-changing urban landscape is the absence of a comprehensive management approach. The most widely accepted response to Jakarta's multiple water challenges is integrated urban water management (IUWM). Actually, IUWM is quite logical since it acknowledges the interconnectedness between the multiple components of the water sector (water supply, sanitation, stormwater, and wastewater), and these need to be aligned with urban development in the larger river basin. The intent of IUWM is to safeguard the quality and quantity of water needed to support the city by integrating all the city water sources with its land-use policy.[25] The obstacles confronting Jakarta to implement some versions of IUWM are especially daunting. There is no system of wastewater management available in vast portions of the urbanized area. The piped clean water network reaches less than 50% of the city's households. There are some general national government guidelines regulating development in proximity of water sources. There is, however, no viable system of regulation of the floodplains that takes into account upstream as well as downstream impacts. Under pressure to serve much of the city's water needs from the underground aquifers, the resulting land subsidence in Jakarta is at the crisis stage.

The current challenges to water management in Jakarta have their roots in the colonial city of Batavia and the decision of building a "water city" in the delta created by the Ciliwung. The approach to managing water inherited from the Dutch colonial system in Batavia remained in force as Jakarta had to accommodate the needs of an emerging megacity. There were warning signs along the way, and multiple recommendations for action to implement an integrated water management system. Not until the era of the great floods beginning in the 1990s did the earlier wake-up calls now sound like the wail of a hurricane warning system alerting everyone that a disaster was coming. First, however, let us examine the water city that the Dutch colonials created.

Notes

1 Yallop, Olivia (2014) "Citarum, the Most Polluted River in the World?" *The Telegraph*, April 11.

2 Soeriaatmadja, Wahyudi (2018) "Military Sent in to Clean Up Indonesia's Citarum River," *Straits Times*, January 10; Valentina, Jessicha (2018) "Government Responds to Documentary Film About Citarum River," *Jakarta Post*, March 1.

3 *Jakarta Post*, January 2, 2019.

4 Argo, Teti A. (1999) *Thirsty Downstream: The Provision of Clean Water in Jakarta*. PhD dissertation, University of British Columbia, Vancouver, Canada.

5 Sardindaningrum, Irene (2017) "Krukut, Jinak di Hilir, Liar di Hulu," *Kompas*, January 23; Wijaya, Callistasia Anggun (2016) "Jakarta Vows to Evict Luxury Houses Along the Krukut River," *Jakarta Post*, September 2.

6 Wijaya, Callistasia Anggun (2018) "What Makes Jakarta's Rivers Ugly and Smelly?" *Jakarta Post*, July 24.

7 Ibid.

8 Lusianawati, Dewi (2014) "Post-Demolishment, Grogol River Dirtied by Trashes," *Berita Jakarta*, September 16.
9 Amira, S., Astono, W., and Hendrawan, D. (2018) "Study of Pollution Effect on Water Quality of Grogol River, DKI Jakarta." 4th International Seminar on Sustainable Urban Development. *IOP Conference Series: Earth and Environmental Science*, 106 (2018): 012023. doi: 10.1088/1755-1315/106/1/012023.
10 Ravesteijn, Wim (2013) "The Tangerang Irrigation Works Prelude (1918-1942) & Interlude (1965-2010)," in Kop, Jan, Ravesteijn, Wim, and Kop, Kasper, eds. *Irrigation Revisited: An Anthology of Indonesian-Dutch Cooperation*. Eburon Vitgeverij BV, pp. 297–316.
11 Blusse, Leonard (2012) "An Insane Administration and Insanitary Town: The Dutch East India Company and Batavia (1619-1799)," in Ross, Robert J. and Telkamp, Gerald J., eds. *Colonial Cities: Essays in Urbanism in a Colonial Context*. Dordrecht, Netherlands: Kluwer Academics; Ravesteijn, op. cit., pp. 307–309.
12 "Pollution Turns East Jakarta River Blue," *Jakarta Post*, May 3, 2019.
13 Verstappen, Herman Theodoor (1953) *Djakarta Bay: A Geomorphological Study on Shoreline Development*. Utrecht, Netherland: University of Utrecht, Drukkerij Trio's Graverhage, pp. 52, 85–88, Figure V, p. 26.
14 Wijaya, Callistasia Anggun (2016) "Jakarta Seeing Results with Rivers," *Jakarta Post*, March 23.
15 Padawngi, Rita and Douglass, Mike (2015) "Water, Water Everywhere: Toward Participatory Solutions to Chronic Urban Flooding in Jakarta," *Pacific Affairs*, 88 (3): 525.
16 Irawan, Dasapta Erwin, Silaen, Henri, Sumintadireja, Prihadi, Lubis, Rachmat Fajar, Brahmantyo, Budi, and Puradimaja, D. J. (2015) "Groundwater – Surface Water Interactions of Ciliwung Riverstreams, Segment Bogor – Jakarta, Indonesia," *Environmental Earth Sciences*, 73 (3): 1295–1302; Cochrane, Joe (2016) "What's Clogging Jakarta's Waterways? You Name It," *New York Times*, October 3. Accessed January 19, 2020: http://nyti.ms/2dy9ykC.
17 Westcoat, James L., Jr. and White, Gilbert L. (2003) *Water for Life: Water Management and Environmental Policy*. Cambridge: Cambridge University Press, p. 218.
18 Pelling, Mark and Blackburn, Sophie, eds. (2013) *Megacities and the Coast: Risk, Resilience and Transformation*. London: Routledge, Chapter 7, p. 201.
19 Djaeni, A, Hobler, M., Schmidt, G., Soekardi, P., and Soefner, B. (1985) "Hydrogeological Investigations in the Greater Jakarta Area/Indonesia." Accessed May 25, 2019: www.swim-site.nl/pdf/swim09/swim09_Djaeni_etal.pdf; Delinom, Robert M. (2007) "Groundwater Management Issues in Greater Jakarta Area, Indonesia," *Proceedings of International Workshop on Integrated Watershed Management for Sustainable Water Use in a Tropical Region, JSPS-DGHE Joint Research Project*, Tsubuka University, October. Bulletin TERC University of Tsubuka, No. 8, Supplement (2): 40–54.
20 Delinom, op. cit.
21 Lubis, Rachmat Fajar (2018) "Urban Hydrogeology in Indonesia: A Highlight From Jakarta," *IOP Conference Series: Earth and Environmental Science*, 118. Accessed May 25, 2019: https://iopscience.iop.org/article/10.1088/1755-1315/118/1/012022.
22 Cochrane, op. cit.
23 Ibid.

24 Van Voorst, Jorgen and Hellman, Roanne (2015) "One Risk Replaces Another," *Asian Journal of Social Science,* 43 (6): 786–810; Leitner, Helga and Sheppard, Eric (2017) "From Kampungs to Condos? Contested Accumulation Through Displacement in Jakarta," *Environment and Planning A: Economy and Space,* 50 (2): 437–456; Dovey, Kim, Cook, Brian, and Achmadi, Amanda (2019) "Contested Riverscapes in Jakarta: Flooding, Forced Eviction and Urban Image," *Space and Polity.* Accessed October 16, 2019: https:doi.org.10.1080/13562576.2019.1667764.

25 Global Water Partnership (2011) *Toward Integrated Urban Water Management.* Stokholm: Global Water Partnership, August.

2 Harnessing the rivers for a water city

The Dutch created the colonial port city of Batavia in 1619 to take advantage of its watery attributes. The benefits and challenges of harnessing and managing this expansive floodplain preoccupied settlers especially during the first two centuries of occupation. As Roosmalen puts it, the

> problem with the supply and disposal of water was something what had been a constant source of concern from the moment the Dutch had first set foot in the colony. Floods during the monsoon, the lack of water during the dry season, a laborious process of securing clean drinking water, and the perpetual problem with the disposal of wastewater were just some of the issues the settlement experienced. It was clear that no improvements would have a lasting effect if nothing was done to improve water management.[1]

Managing water was basic to the inner workings of the city, much like in homeland cities such as Amsterdam. Keeping out the sea to prevent flooding, using the sea to flush the settlement of wastes in its canals, constructing waterways to drain land to create additional habitable spaces, supporting the trade functions of the port, providing a familiar urban aesthetic to compensate for the difficulties of foreigners living in a strange new environment, and, of course, ensuring a sufficient fresh potable water supply to support the needs of the population, all were parts of the complex system of water management central to effectively governing the city from its beginnings. The fact that Batavia perched on slightly elevated ground surrounded by marshlands on all sides is noteworthy. Thomas S. Raffles' *History of Java* affords a detailed, firsthand account of the city's site as he saw it in the early 19th century, even after nearly two centuries of settlement and a continuous process of land improvements.

> Batavia was built almost in a swamp, surrounded by marshes in all directions ... Opposite the mouth of the river [on which it was located], and extending a great way to the westward, is a mud-bank, which in many parts at low water is uncovered by the sea, and is daily accumulating from quantities of mud and animal and vegetable matter carried down by the river during its reflux. Again, the sea often at spring tides overflows the adjacent country, and on

DOI: 10.4324/9781003171324-3

its receding leaves the soil covered with slime and mud, which are carried by the sea breeze over Batavia.[2]

Raffles gives us another portrait of the city's early waterways, one that underscores the difficulty of keeping these urban thoroughfares clear of debris. As he noted, "the stagnant water of the canals, which in all directions intersect the city are filled with filth of every description." These conditions did not prevail in the settlements growing up outside the old city. "Weltevreden, at a distance of not more than three miles, being less exposed to these causes, excepting the water, is exempt, in a great measure, from its prevailing endemic fever."[3]

Already on the site when the Dutch arrived was a small settlement, known as Jayakarta, which had for centuries served as the sea access for the Sundanese kingdom that was based inland near what is now the city of Bogor. The port at Jayakarta was Sunda Kelapa. This modest settlement situated along the west bank of the Ciliwung relied upon the existing natural water landscape both as the connector to the sea and as the primary source of its drinking water. Its new ruler, the Sultan of Banten renamed it Jayakarta early in the 16th century. At its peak in the late 17th century, Banten was the largest port in West Java with a population, including its surrounding villages, of nearly 700,000. It rivaled the Portuguese-controlled port of Malacca on the Malaysian peninsula for trade dominance in the region. In the late 16th century, according to historian Anthony Reid, Banten was one of the region's most cosmopolitan places. Sultan Abdulkadir, who reigned from 1596 to 1624, allowed first the Dutch in 1596 and then the English in 1602 to establish trade facilities in Jayakarta.[4]

A third European rival, the Portuguese, did not fare as well in Jayakarta under the Sultan's rule. Prior to its seizure by Banten, the Portuguese had established a foothold in Sunda Kelapa with permission from the Sundanese. When it was attacked and conquered by the Banten military leader Fatahillah in 1526, the Portuguese lost their foothold there and were denied permission to construct a fort to sustain their trading position. In fact, the transfer of power from the Sundanese to the powerful city of Banten provided the impetus for both the British and the Dutch to usurp the trade monopoly there that the Portuguese previously enjoyed. The Banten Sultan constrained foreign influence in Jayakarta, and this prevented the settlement expanding as it might have under a different political arrangement. Throughout the 16th century, Banten kept Jayakarta's development in check. Even so, on the eve of the Dutch takeover in 1619, it boasted approximately 10,000 inhabitants in the town and within its sphere of influence.[5]

Jan Pieterszoon Coen, Governor-General of the Dutch East India Company (VOC), orchestrated the seizure of Jayakarta from Banten. His plan was to establish a network of Dutch ports throughout the East Indies to compete with England and Portugal, as well as to overcome opposition at home to foreign interventions. Founded in 1602, the Netherlands government granted broad powers to the Dutch East India Company to function as its surrogate within Asia, including the power to wage war and to conclude treaties on behalf of the home country.[6] During the 17th century, they operated a trade network stretching from India eastward to Japan, including a massive fort at Tayouan (on Formosa), with their ships carrying on commerce throughout the region.[7] Ultimately, the VOC concentrated its control of Asian trade from strategic sites in the Indonesian archipelago. Given the open hostility toward the Dutch by the Sultan of Banten and perpetual friction with the English shippers who had been allowed to establish a warehouse facility in Banten, Coen saw the advantage of shifting the Dutch operations from Banten eastward to Jayakarta, a location still within easy access of the navigable Sunda Straits and beyond the Sultan's close control. The VOC secured permission from the Prince of Jayakarta to build a fort across from the town (on the eastern side of the Ciliwung River). The British had a warehouse facility adjacent to the main indigenous settlement on the western side of the river. Late in 1618, the Dutch and English began to spar over their respective roles in Jayakarta, leading to an unsuccessful siege of the Dutch fort by the British East India Company forces. Not wanting the Dutch to gain an upper hand in this struggle and not trusting the Prince of Jayakarta to represent Banten's interests during this conflict, the Sultan sent his own fleet to bolster the strength of the British and to ensure victory by the British. The result was a three-month siege of the Dutch fort but with no successful attack. Given the disappearance of a military threat owing to the lack of aggression by the English and the Jayakarta Prince (who in the meantime had been removed from his position by the Banten Sultan for the lack of military action), the Dutch forces claimed the lull as a victory. On March 12, 1619, the Dutch celebrated the resiliency of their stronghold by naming it Fort Batavia, and simultaneously claiming sovereignty over the eastern bank of the Ciliwung. Coen, who had gone to get additional forces from a Dutch fort in the eastern island of Ambon during the siege, returned in May with more forces. With these additional forces, he seized the Indonesian town on the other side of Ciliwung, destroyed the Sultan's court, and declared Batavia (now encompassing both sides of the Ciliwung) the new seat of the Dutch East India Company. Although the Dutch in Batavia faced attacks from Banten in 1628 and 1629, the strength of their naval forces, coupled with the internal strife among the Javanese, enabled the Hollanders to consolidate their hold on Batavia. Except for occasional marauders from Banten, Batavia experienced no serious military threats after 1629, the year Coen died. When the Sultanate of Banten acquiesced to Dutch dominance in 1684, it became possible to expand settlement at Batavia beyond the walls of the city.[8] In the meantime, and no longer threatened by external forces, Batavia's leadership directed their energies to harnessing the surrounding waterways to construct the larger community beyond the walls of the fort.

Managing the colonial waters

Historian Leonard Blusse offers a detailed portrait of how Batavia engineered this watery environ into a replica of a Dutch port over the course of the 16th and 17th centuries. As he notes, after unsuccessful sieges by Javanese armies in 1628 and 1629, with the Dutch in seeming permanent control, considerable efforts were expended to construct a system of canals to manage the waterways, a process that includes canalizing the Ciliwung to support the town's development and to ensure easy access to the growing shipping traffic. Under Governor-General Jacques Specx, who replaced Coen, Chinese laborers secured through the Chinese contractor, Jan Con, constructed a new canal on the western side of Batavia castle. More than 2,000 Chinese laborers worked on this and other public works projects. Specx's short stint as governor-general that ended in 1632 proved highly productive. The Chinese laborers, grateful for the work his administration provided, fabricated a commemorative medal in Specx's honor. The medal depicts the completed network of canals and managed rivers that the governor-general had authorized and were built, including the new canals on the western side of the Ciliwung, where the court of the Prince previously stood. This area now supported food production within the city's fortifications. As the local European leadership recognized quickly, the Chinese were indispensable when it came to building and managing public works. As Henrick Brouwer wrote to the company in Amsterdam,

> None of the Dutch burghers is willing to contract for building projects such as the dredging of canals or the supply of wood, lime and stone. Only Chinese are engaged in this sector. Without their help the construction of Batavia's fortifications and the city's present lay-out would have required many more years to complete.[9]

In 1638, following the wet monsoon, the local leadership again contracted with Jan Con, this time "to replace the wooden palisades along the western, northern, and southern edges of the town with stone city walls which would require less maintenance and could provide more security," Blusse notes.[10]

As evidenced in available maps depicting the ground plan of Batavia,

> the entire town was surrounded on all sides by water: canals on the east, south, and west: the Java Sea to the north. Also on the north was a fort surrounded by water and walls with its eastern side projecting beyond the shoreline.[11]

The main streets of the 17th-century city were a combination of natural waterways and engineered canals fed by the Ciliwung (Figure 2.1). The engineering involved straightening the Ciliwung to function like a canal. As Caljouw, Nas, and Pratiwo note, the system of canals dug during the early 17th century were "very much alike that of Amsterdam at the time, while the reclaimed ground was used to raise the land for construction purposes."[12] The construction of a city that drew inspiration from the familiar patterns of settlement at home, at least initially, compensated somewhat for the difficulties of living so far from the Netherlands in a place with a radically different climate. As Abeyasekere put it, "the Company ... dug a system of canals to surround and penetrate the town, giving it a typically Dutch appearance."[13] The city's shipyards replaced the agricultural plots located

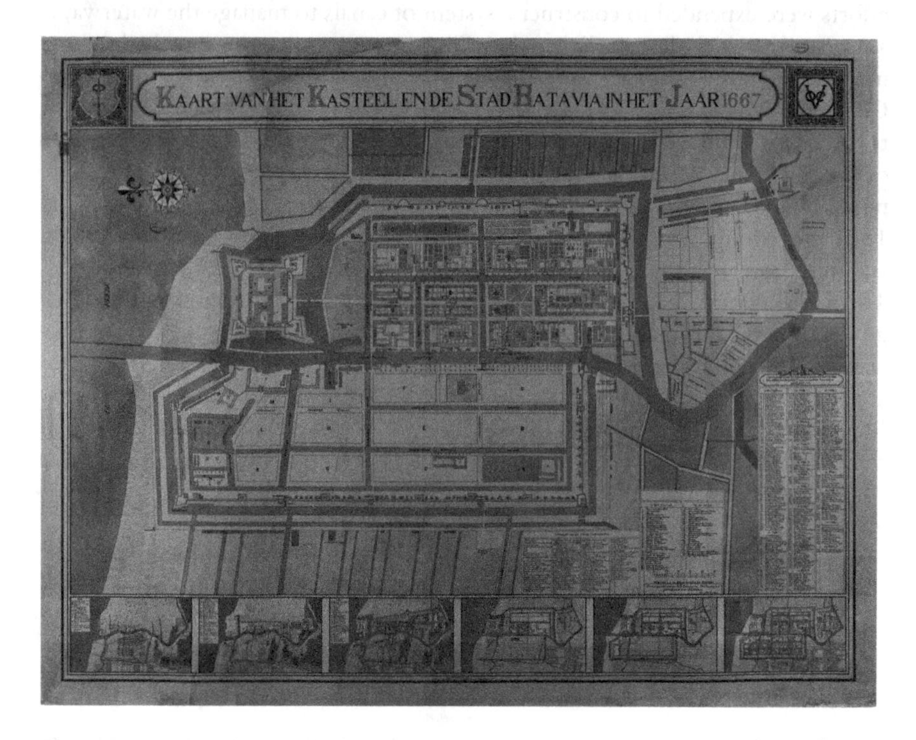

Figure 2.1 Castle and town of Batavia in 1667 showing the canal system.
Source: Collection of the Nationaal Museum van Wereldculturen. Coll. no. TM-496-2.

on the western side of the Ciliwung and offered space for workingmens homes and an area set aside for the Chinese settlement. Outside the city walls to the south along the Ciliwung and separated by another canal were the sparsely settled southern suburbs. So long as the main settled areas remained within the confines of the wall, the system of engineered waterways served as the arteries that bound together the community.[14] According to historian de Haan, the "Dutch changed the natural landscape to fit their conceptions of town planning, deepened and straightened the rivulets, and moved the mud from the channels onto adjacent land to increase elevation."[15]

According to testimony provided by visitors to Batavia, these engineered waterways afforded the city an elegance on par with those in the homeland. The Dutch visitor Valentijn in 1726 remarked on Batavia's "elegantly planted straight canal" (the Tijgergracht) with its "uniformly beautiful buildings," claiming that it was as fine as any of the great canals in Amsterdam.[16] In addition to the elegant Tijgergracht, two other principal north-south streets, the Heerestraat and the Princenstraat, possessed elegant structures. They refashioned Ciliwung rivulets into one new street, Groenestraat, and three new canals, the Groenegracht, Amsterdamsegracht, and Leeuwinnengracht.[17] After 1639, Batavia dug two

Figure 2.2 Working canal in Batavia.
Source: Collection of the Nationaal Museum van Wereldculturen. Coll. no. TM-10014881.

more canals on the western side of the Ciliwung (the Maleischegracht and the Rhinocerogracht). Additional canals appeared later in the 17th century as settlement on the western side of the Ciliwung actually pushed northward beyond the fort on the eastern side. A sharp curve in the Ciliwung south of the city walls created the space where Batavia's Chinese community resided.[18] Figure 2.2 shows how these canals remained as active commercial avenues in the Chinese community into the early 20th century.

Altogether, the city Valentijn visited in the early 18th century had 8 streets, 16 canals, and 56 bridges. "The canals were lined with beautiful vegetation that gave fragrance and shade throughout the day."[19] There were occasional problems with flooding, particularly during the Spring tides. More challenging and more pervasive, however, was when the rains did not come or when agricultural needs usurped the supply intended for the city. Under these circumstances, the canals dried up, exposing the debris that could not get flushed out and subjecting the settlement to health and logistical challenges.

Over the course of the 18th century, however, the grandeur of Batavia faded, not just because of the economic conditions leading to demise of the Dutch East India Company in 1799 but because of, as Blusse contends, the steadily deteriorating environmental conditions, especially regarding the critical water supply.

Another factor was a massive urban reconstruction project that began a process of expanding the city beyond its original core area. A combination of inadequate local planning, reduced maintenance, and a misguided decision to rely upon sugar plantations as the main export crop of the region contributed to the eroded supply and quality of water that flowed into Batavia. Blusse drew upon a report by Couperus, who visited Batavia in 1815, to document the deteriorated state of the city's famed canal system. Couperus found the famed Tijgergracht to be a shell of its former grandeur. As he wrote,

> this is the formerly beautiful Tijgergracht, but all the manifold edifices which used to occupy the area in between this canal and Prinsenstraat have been levelled (sic) to the ground. Who would recognize the Tijgergracht which Valentijn celebrated in 1726, only eighty years earlier?[20]

What Couperus witnessed was the aftermath of the restructuring of the city, launched by Governor-General Daendals after 1800. He tore down the old city walls, filled in many of the canals, and shifted the Batavia's administrative and the European residential functions to the "higher lying regions" of Rijswijk, Noordwijk, and Weltevreden, to the south of the original settlement. Material from the city walls became the stone for a new governor-general palace constructed in Weltevreden. Daendals' aggressive redevelopment efforts have been widely documented in the historical literature and often tied to the seemingly pragmatic and cost-efficient strategy of using the materials from the old city walls to erect the new public facilities being constructed in Weltevreden. The filling in of much of the canal system was a response to the long-recognized problems of pollution in the old city that were widely believed to be associated with the high mortality rate for those residing within the city walls.[21]

How the waters in the canals became polluted was a point of dispute. One commonly held view was that the density of habitation along the edges of waterways, coupled with faulty hygienic practices, especially the practice of dumping household waste directly into the river, were the key factors. It was an accepted theory up through the late 19th century that the evil gases associated with the debris (the so-called miasmas) caused diseases, so moving the population away from the canals, or filling them in, seemed a wise and necessary health precaution. The polluted conditions in the canals, according to most accounts, explained why the European community moved inland to the higher ground to the south Rijswijk and Noordwijk, and further away in Weltevreden.[22]

According to Blusse, however, it was not just faulty hygiene that explains the pollution of the waterways in the old city. He points to the decision of the Dutch East India Company beginning in the 1630s to emphasize large-scale cane sugar cultivation in the hinterland of Batavia that affected the quantity and quality of water available to the settlement. The process of sugar cultivation required draining the land, establishing an irrigation system through a canal network, and utilizing vast quantities of wood to process the raw material. These processes quickly led to denigration of the surrounding landscape. Sugar cultivation devoured the trees, polluted the water with waste material from the cane, and quickly exhausted the soil (which also contributed to erosion and land subsidence) in what previously was a heavily forested area that served to recharge the aquifer. By the end of the 17th century, city's leadership recognized a sharp decrease in the volume of water coming into Batavia from the vast river system that flowed from the mountains to the south. Even with the regular dredging of the canal system in the settlement area by the Chinese laborers, there was a continuing problem of reduced flow even during the wet season, and reduced flows increased the problem of stagnation of water in the canals. Management of canals, as well as the bridges and roads, was one of the key concerns of the local leadership. Since 1664, the responsibility for managing the waterways existed under the authority of a board of *dike reeve*, similar to its counterpart in the Netherlands. The *dike reeve* exercised power to require local inhabitants to contribute to maintaining the bridge and canal facilities.[23] Overall, less maintenance occurred once the European settlement moved southward toward Weltevreden.

Following the Mt. Salak volcano eruption in 1699, it seemed that the water supply slowed even more. In 1701, the Batavia government sent a reconnaissance party into the interior to find out if that volcanic eruption affected the rivers and accounted for reducing the flow and increased pollution. They found little evidence that the volcanic eruption contributed to the problem, but solid evidence that the diversion of water to irrigate the sugar cane plantations was the culprit and accounted for the reduced water flow and accelerated pollution in the Tangerang River. At the time, the waters of Ciliwung seemed much clearer but over the next several decades its flow dropped precipitously. By 1720, there was a chronic water shortage in Batavia, and the rivers that now flowed much more slowly through the city seemed more polluted. Owing to the buildup of silt (and the less investment in regular dredging), a sandbar formed at the mouth of the Ciliwung making commerce more difficult and impeding the flushing of the canals.[24]

There were efforts undertaken to reduce the problems of silting, to mitigate pollution, and to control flooding during the rainy season. The city constructed a dam in 1725 that added to the Ciliwung a flood control channel. Frequently the volume of water during the wet season overwhelmed this facility, with the result that the river breached the engineered walls and returned temporarily to its natural course. An expanding system of canals outside the main settlement diverted river water to support the agricultural areas and attempted to control the flow of water which passed through the city during the rainy season to reduce the threat of flooding. The Westerse Vaart "was the first in a series of flood canals aimed at controlling flooding," followed by the Bacherachtsgracht and the Mookervaart which directed water coming down from Bogor westward around the settlement.[25] The Mookervaart Canal was constructed in 1732 under Governor-General Diederik Durven to bring a larger water supply to the Batavia settlement. Owing to faulty engineering, the Mookervaart Canal became a stagnant pool rather than a flowing waterway. It exacerbated the problem of pollution in Batavia and contributed further to the silting problem that clogged the city's transportation arteries. As Blusse puts it, "the inept construction of the Mookervaart, meant to alleviate the sufferings caused by a decrease in water supply, paradoxically resulted in the destruction of the ecology of the town's direct surroundings."[26]

The canal and river systems were major contributors to the unhealthy conditions in the old city. The available data suggest that Batavia's already high mortality rate increased further after 1720. This encouraged outmigration of residents to the new suburban region further upriver and on higher ground. In fact, conditions had become so decrepit in Batavia that the Dutch East India Company seriously considered moving its headquarters to another port city, either Semarang in Central Java or Surabaya in East Java. Instead of such a drastic move, however, under Governor-General van Overstraten in 1797, the administration of Netherlands Indies (which had taken over colony from the bankrupt Dutch East India Company) began to move the government from its coastal site to nearby Weltevreden south of the city. Governor-General Daendals completed the transition, using the material from the castle, the walls, and the outer fortresses to build new structures in Weltevreden. As previously noted, Daendals filled in many of the canals in an effort to put an end, so he said, to the scourge of fetid pools.[27]

The movement of the European community to Weltevreden, completed during the 19th and early 20th centuries, eliminated public pressures to address the problems of water pollution and flooding in the original sections of the city. During the administration of Thomas Raffles between 1811 and 1816 (coinciding with the British interlude that followed the French era under Daendals), the city of Batavia was a different place. As he noted in his *History of Java*, little was left of

> the splendour and magnificence which procured for this capital the title of the Queen of the East ... Streets have been pulled down, canals half filled up, forts demolished, and palaces leveled with the dust. The stad-house, where the supreme court of justice and magistry still assemble, remains; merchants transact their business in the town during the day, and its warehouses still contain the richest productions of the island, but few Europeans of respectability sleep within its limits.[28]

For his part, Raffles rejected the luxury of the new Weltevreden-based governor-general palace for his residence and chose to conduct the affairs of state from the higher and drying elevations of Bogor more than 60 kilometers to the south. The abandonment of the old city by the Europeans did not mean that all departed. Data show 47,217 people still resided within city and suburbs up to two miles to the south in the early 1800s. This was broken down as follows: 543 Europeans; 1,485 descendants of Europeans; 11,249 Chinese, 14,239 slaves, and 9,749 Makassar/Bugis; 3,331 Javans, and 3,155 Malayus. Those non-Europeans who remained within the original settlement experienced routinely the problems of pollution, flooding, inadequate water, and the associated health problems. Suburbanization, on the other hand, afforded Europeans some relief and an appreciable decline in the mortality rate.

According to data provided by Abeyasekere (1987), the migration of the European community away from the original settlement brought a sharp drop in its mortality rate from 228 to 53 persons per 1,000 between 1819 and 1844. By the 1900s, it dropped further to 29 persons per 1,000. This improved life expectancy bypassed the Indonesian and Chinese kampungs, however. The Chinese moved into the old city from the former Chinese camp in Glodok (outside the city walls when they were still standing) to take up residence in areas abandoned by the Europeans. Mortality data covering 1903–1911 indicate that the death rate for the Chinese was 45 person per 1,000 and for the Indonesians it was 64.3 per 1,000, a rate more than double that of the Europeans.[29] These crowded areas in the old city and at the fringes where the Indonesians resided routinely experienced flooding, with waters sometimes rising to three or four feet in their humble dwellings. These Indonesian kampungs were located throughout the settled areas but typically on less desirable and "often swampy land." For those Indonesians who resided on the higher ground south of the old city, the death rate of 48 person per 1,000 was half that of the those living in the older city at 98 persons per 1,000.[30]

John Stockdale's portrait of Batavia published in 1811 suggests that problems with what remained of the canal system persisted. He noted that the city suffered from stagnant water in the dry season and flooding in the lower part of the town during the rainy season, rendering "Batavia one of the most unwholesome spots of the face of the globe."[31] Part of the problem, he noted, was a sandbar that formed at the mouth of the "Jaccatra River" (this was the Ciliwung) that prevented ships from entering even though the river had a width of between 160 and 180 feet.[32] Stockdale's description is worth presenting in full in order to appreciate what he regarded as the key factors explaining the "insalubrity" of the old town in the early 19th century. It also points to the implications of misguided efforts to prevent flooding that had even worse consequences.

> [T]he little circulation of water in the canals which intersect it [the town] and the dimunition of its inhabitants ... is occasioned by the river, which formerly conveyed most of its water to the city, being now greatly weakened by the drain which has been dug, called the *Slokhaan*, which receives its water from the high land and carries it away from the city, so that many of the canals run almost dry in the good monsoon. The stagnant canals in the dry season exhale an intolerable stench, and the trees planned along them impede the course of the air, by which in some degree the putrid effluvia would be dissipated.

Figure 2.3 shows the Molenvliet canal that relied upon the river system to maintain its viability as a water source for the Batavia communities. In Stockdale's view, flood control and water diversion engineering undertaken during the dry season failed to rectify the circumstances in the wet season. As he continued,

> In the wet season the inconvenience is equal; for these reservoirs of corrupted water overflow their banks in the lower part of the town, and fill the lower

Figure 2.3 Molenvliet canal passing through Rijswijk and Noordwijk in Batavia.
Source: Collection of the Nationaal Museum van Wereldculturen. Coll. no. TM-60027089.

> stories of houses, where they leave behind them an inconceivable quantity of slime and filth; yet these canals are sometimes cleaned, but cleaning of them is so managed as to become as great a nuisance as the foulness of the water, for the black mud taken from the bottom is suffered to lie on the banks, in the middle of the street … as this mud consists chiefly of human ordure.[33]

Throughout the 19th century, the problems of pollution in canals, inadequate drinking water during droughts, as well as flooding during the rainy season challenged Batavia's leadership. At the end of the century, faced with the problem of insufficient clean water, the city shifted from surface to sub-surface water sources by constructing several new artesian wells. Between 1873 and 1877, Batavia added seven artesian wells. The additional wells and accompanying reservoirs increased the volume of available hygienic water.[34] The groundwater served principally the European community, although native areas derived some benefit from the new water. Two problems hampered efforts to improve clean water access to the Indonesian kampungs. One was simply that the indigenous community regarded the water from the artesian wells as

Figure 2.4 Townscape of Batavia in early 1900s.
Source: Collection of the Nationaal Museum van Wereldculturen. Coll. no. TM-10014875.

inferior to the river water, and thus, many subjected themselves unwittingly to the more polluted source rather than use the public taps. As Figure 2.4 shows, the active commercial activities in the rivers contributed to the pollution. The other problem was that the public taps often failed due to lack of maintenance. In 1888, the water hydrants in two old town kampungs broke, requiring inhabitants to purchase expensive water from merchants or continue to use the polluted river water.

It was not just limited access to clean water that differentiated the experiences of the indigenous kampungs and the European settlements. Flooding "happened so frequently" in the kampungs that it was accepted as a normal condition of life in Batavia. "Only when there were particularly bad floods, as in the 1870s, did the authorities feel obliged to undertake large drainage works." In particular, the flood of 1873 covered the entire city with at least one meter of water. Following the 1873 flood, the government expanded the drainage canals dug in the 1820s and filled in some of the older canals in the lower city to avoid the problem of stagnant water. These were the canals filled with silt that no longer possessed a self-flushing capacity. Abeyasekere contends that "in general, Chinese and Indonesian in the lower town suffered water-related problems – either too much of it or too little, polluted or in the wrong place."[35]

Yet according to Augusta de Wit, a European born in Sumatra and who taught in a girl's school in Batavia, the ambiance of the *waterstad* (water city) was still to be found in the new fashioned upper town, "that picturesque ensemble of villa-studded parks and avenues." The new settlements were unlike the older Batavia, grey, grim and stony as any war-scarred city of Europe – the stronghold, which the steel-clad colonists of 1620 built on the ruins of burnt-down Jacatra. But, long since abandoned by soldiers and peaceful citizens alike, and its once stately mansions degraded to offices and warehouses, it has sunk into a mere suburb – the business quarter of Batavia – alive during a few hours of the day only, and sinking back into death-like stillness, as soon as the rumble of the last down-train has died away among its echoing streets. And the real Batavia – in contradistinction to which this ancient quarter is called "the town" – is as unlike it as if it had been built by a different order of beings.[36]

De Wit's description of the canal functions in the newer quarters of Batavia, including the areas of Rijswijk, Noordwijk, and Molenvliet, captures how life along the canals retained some of its traditional charm as the city entered the beginnings of the modern era, a transition reflected in the growing role of streetscars to move the city's population. It is worth citing the text in full:

> Rijswijk, Noordwijk and Molenvliet, the commercial quarters of Batavia, are more European in aspects than the Koningsplein; the houses – shops for the most part – are built in straight rows; a pavement borders the streets, and a noisy little steam-car pants and rattles past from morning till night. But, with these European traits, Javanese characteristics mingle, and resulting effect is a most curious one, somewhat bewildering withal to the new-comer in its mixture of the unknown with the familiar. Although commonplace-shops are approached through gardens, the pavement is strewn with flowers of the flame-of-the-forest: and at the street corners, instead of cabs, one finds the nondescript sadoo, its driver, gay in a flowered muslin vest and a gaudy headkerchief, squatting cross-legged on the back seat. Noordwijk is unique, an Amersterdam-"gracht" in a tropical setting. Imagine a long straight canal, a gleam of green-brown water between walls of reddish masonry – spanned from place to place by a bridge, and shaded by the softly-tinted leafage of tamarinds; on either side a wide, dusty road, arid gardens, sweltering in the sun, and glaring white bungalows; the fiery blue of the tropical sky over it all. Gaudily-painted "praos" glide down the dark canal; native women pass up and down the flight of stone steps that climbs from the water's edge to the street, a flower stuck into their gleaming hair, still wet from the bath; a tribe of fruit vendors and sellers or sweet drinks and cakes have established themselves along the parapet, in the shade of the tamarinds; and the native crowd, coming and going all day long, makes a kaleidoscopic play of colours along the still dark water. From the little station at the corner of Noordwijk and Molenvliet, a steam-car runs along the canal down to the suburbs; every quarter of an hour it comes puffing and rattling; and every time the third-class compartment is choking full of natives ... The skippers and raftsmen are more conservative in their ways – owing, perhaps, to their constant communication with the deliberate stream, which saunters along on its way from

Figure 2.5 Women washing along the Molenvliet.
Source: Collection of the Nationaal Museum van Wereldculturen. Coll. no. TM-60005949.

> the hills to the sea, at its own pace. They take life easily; paddling along over the shifting shallows and mud-banks of the Kali (river) in the same leisurely way their forebearers did; conveying red tiles, bricks and earthenware in flat-bottomed boats; or pushing along rafts of bamboo stems, which they have in the wood up-stream ... For keep house they do in their boats. They eat, drink, sleep, and live in the prao.[37]

Adding to the character of life along Batavia's rivers was how they served as the place to conduct household chores not possible in kampung communities that lacked access of water services (Figure 2.5).

Modernizing the colonial city

A widely read and critical appraisal of colonial life, *Het Leven in Nederlandsch Indie* (The Life in the Netherlands Indies) written in 1900 by Bath Veth, expressed the view that water was the great challenge in the colonies, "polluted, dripping, leaking, or flowing unregulated." In the 17th and early 18th centuries, managing water was largely associated with supporting irrigation of agricultural areas as well as meeting the needs of the small settlement. As shown in Figure 2.6,

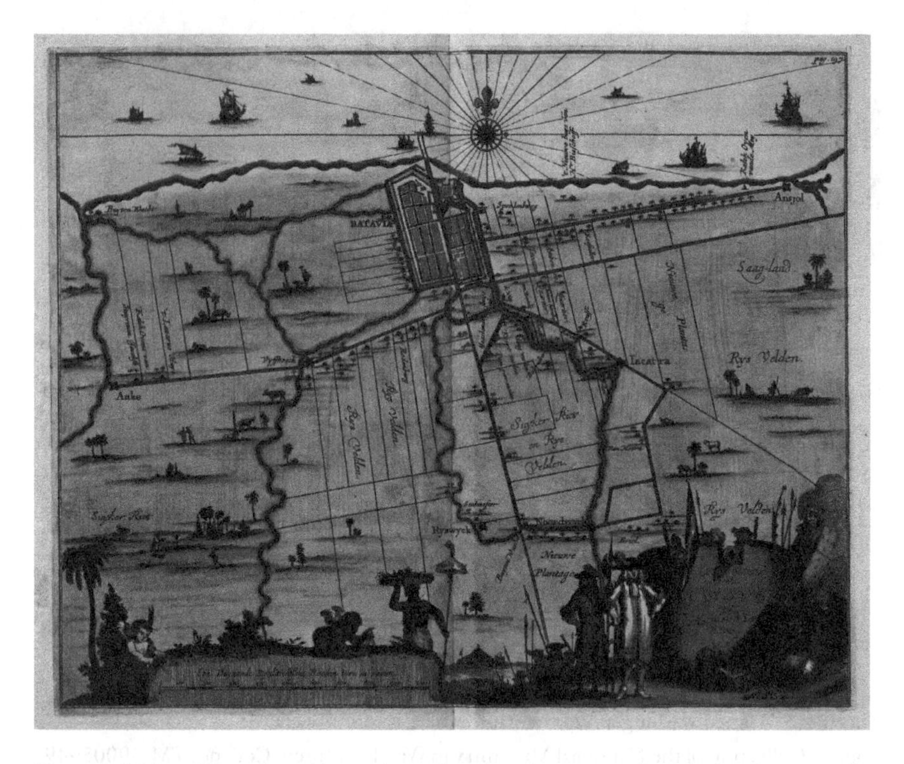

Figure 2.6 Map of Batavia and its environs showing the system of canals drawing upon the river system to support agriculture.

Source: Atlas of Mutual Heritage MH-5641 KB.

there was an extensive system of canals that drew water from the major rivers before it reached Batavia.

By the late 19th century, as Mrazek puts it, it had become almost exclusively "a problem of urbanization." Popular magazines referred to "pipelines for drinking water in the Indies cities among the cornerstones of colonial power."[38] Addressing urban water needs for the expanding European population was proof of the efficacy of imperial rule. Mrazek notes that as the Dutch authorities devoted more attention to fashioning the modern 20th-century colonial city, it gave increased attention to the provision of sufficient safe water and the removal of waste.[39] This included reconstructing the port facilities that had been Batavia's *raison d'etre* from its inception.

The construction of a modern port at Tanjung Priok began in 1877 and took nine years to complete. The inaccessibility of the traditional harbor at Sunda Kelapa because of the expanding sandbanks that blocked the waterways made landing goods there increasingly difficult and expensive. When completed, Tanjung Priok gave Java its first modern deepwater harbor that allowed large ships to tie up directly at the wharf, rather than anchor out in the bay and transfer goods by small boats. One reason for the port's protracted development process was the battle over its location. Local merchants led by the Batavia Chamber of Commerce favored building it just to the west of Sunda Kelapa. This necessitated constructing a dam connecting the shoreline to nearby Onrust Island that served as Batavia's ship repair facility. As Veering described it, the "dam would provide shelter from the wind and waves for boats anchored there whilst at the same time serving as a loading and unloading quay."[40]

This, however, would put the harbor nearly 20 kilometers from Batavia and require a rail line to ship the freight and passengers to the city. The Netherlands East Indian Railway Company conceived and favored this plan since it extended their franchise. The director of the Public Works Administration, W.E.H.F. van Radar, offered an alternative scheme, locating the harbor nine kilometers to the east of Batavia at Tanjung Priok at the mouth of the Sunter River. Although the government initially authorized funding for the western harbor, the Dutch Lower House in the Netherlands rejected it and called for fuller consideration of other alternatives. One of the fears of the Batavia merchants with the Tanjung Priok site was the chance that it would shift all commercial activities to that site and become, in fact, a new city to rival Batavia. Opponents of the Tanjung Priok site could not have seriously thought that a new city would emerge there since the area was a large swamp. A final report by the Batavia leadership in 1875 had the majority favoring the western site but a small group pressed the Tanjung Priok option. A team of analysts headed by two prominent hydraulic engineers, who had worked on harbors in Amsterdam and Rotterdam respectively, reexamined the options. They favored the Tanjung Priok option because there they could construct a facility deep enough and equipped with modern facilities, whereas the western site plan extended the shallower Sunda Kelapa. The opinion of the engineers carried the day. The dredging and construction of the quays took place between 1878 and 1883. The eruption of the volcano Krakatoa on August 27, 1883, which generated a plume of ash seen as far away as England, caused some damage to the Tanjung Priok jetties because of massive tidal waves. The new harbor proved its mettle, however, and survived virtually unscathed. When finally completed in 1885, it represented the most ambitious, modern, and successful public works project in the Netherlands Indies and continues to serve Jakarta as its main commercial harbor.[41]

A new canal connected Tanjung Priok to the old town, a project that like previous canal excavations took a heavy toll on the laborers consigned to the task. Except for some of the harbor laborers, however, no one took up residence near Tanjung Priok. The threat of a rival city rising up at Tanjung Priok never materialized. Visitors arriving at the harbor by ship typically took the train into town and left the cargo to move on the slower canal route.[42] With a new deepwater harbor far from the traditional location, and the European population departing to the higher grounds in Weltevreden and serviced by water from wells and far from the remaining polluted canals, water management in the rivers and canals no longer seemed a pressing problem. The poor water quality of the "lower" city, the need for regular dredging, and the annoyances of flooding no longer troubled the

Europeans. For the indigenous communities in the kampungs and Chinese community that moved into the old city, however, nothing changed. The problems persisted but without any remedies.

The growth of the European community in the new "upper city" necessitated expansion of the water system, however. According to Jan Kop, to serve the clean water needs of the main European community, it abandoned completely its former reliance on water drawn from upstream on the Ciliwung and turned exclusively to artisan well water complemented by the addition of spring water sources. By 1920, Batavia boasted 28 artisan wells, 12 reservoirs, and a distribution system providing household connections throughout much of the European community. Spring water was carried to the settlement through a 55-kilometer pipe system from the highlands at Bogor. Set up in 1923 the spring water supply. Increased the system capacity to 350 liters per second and expanded the reservoir size from 780 to 20,000 cubic meters. The water network extended 150 kilometers through the urban area. For a brief period, the native population had access to water hydrants at no charge. Yet even with these expanded water sources, there was concern that continued population growth would soon exceed the capacity of the wells and the spring water. One proposed remedy called for transporting water from the Ciliwung to the West Flood Canal to be purified as an additional source. Research on ways to improve water supplies conducted at the Institute of Technology in Bandung through its Water Purification Experimental Station studied this plan. Research on prevention of diseases owing to poor water quality also began. Opened in Batavia in 1927, a new Medical High School focused on prevention and alleviation of "water-related diseases."[43] One problem with returning to using river water, by drawing from the West Flood Canal, was the possibility of introducing inferior water into Batavia's system. Even the native settlements that still relied on river raised concerns among the city's health inspectors who recognized the link between clean water and a healthy city. As one inspector noted,

> The need to care for good drinking water is being recognized more and more. People begin to see that carefully draining off the feces may well pay back in reducing the occurrence of infectious illnesses.[44]

Besides improved drinking water service to support the European community, the government considered ways to ensure that the problems of flooding in the lower city did not become endemic to the communities moving inland near some of the major rivers. Herman van Breen, an engineer and member of Batavia's town council, devised a plan to improve drainage and enhance the supply of water to prevent flooding, to regulate local drainage to reduce water pollution, and to remove continuing hazards to shipping. His 1917 plan called for a flood canal to protect the newly built homes of Europeans in the Menteng community. The flood canal would intercept upstream water before it reached the city and divert it to the east and west in a semi-circular trench that would carry it into the Jakarta Bay.[45] The van Breen flood canal would be the first serious canal project in Batavia in more than a century and if fully implemented would be very costly. The first major flood in more than 50 years occurred in 1918, and the timing of this flood certainly contributed to a new sense of urgency to undertake something to protect the upper city settlements. It took another four years, however, to begin construction on the flood canal. The plan called for two floodgates, one at Manggarai and a second at Karet. These were in addition to the sluice gates (see Figure 2.7) that regulated water flows within the European community during normal conditions. Where the project departed from the original van Breen plan was the decision not to construct the eastern section of the canal. It would be nearly another century, and many floods later, before the city constructed the eastern section. Yet even when it did get constructed, many objected, not so much because of the cost, but that by the early 21st century such a small portion of the "inner" city would be protected by a flood canal devised to protect the city that had grown five times larger than it was in the 1920s.[46]

The van Breen plan coincided with other initiatives from Batavia's political leadership aimed at improving the overall living conditions in the colonial city. Several factors explain this activism, one being the new overall colonial policy framework emanating from the Netherlands. A speech by Queen Wilhelmina in 1901

Figure 2.7 Sluis gate at Noordwijk in the European settlement inland from the original settlement.

Source: Collection of the Nationaal Museum van Wereldculturen. Coll. no. TM-60005550.

set forth what was commonly referred to as the Ethical Policy. She proclaimed that the Dutch owed their colonies a "debt of honor." The Indies gave much to the mother country over three centuries, without the Dutch reciprocating in a commensurate fashion. The Ethical Policy sought to redress this neglect and set in motion initiatives to improve conditions for the indigenous populations. As historian Frances Gouda put it, the Dutch intended to govern the Netherlands Indies with greater sensitivity to the needs of the indigenous populations than other European colonial powers governed their territories. The Ethical Policy coincided with a move to decentralize decision-making from the Netherlands to Batavia, thereby expanding considerably the functions and responsibilities of local government. The Decentralization Law of 1903 bestowed power to municipal corporations in Java to develop, oversee, and maintain roads, parks, cemeteries, markets, and slaughterhouses and to be in charge of firefighting, public health, transport, street lighting, housing, water, waste disposal, and even site preparation for developing new urban areas.[47] With a growing European population now regarding Batavia as their permanent home rather than just a stopover on the way to building a career, there was an impetus to make improvements of a permanent nature. Despite the sentiments intended by the Queen's Ethical Policy, much of the effort benefitted the European community with only marginal investments in the indigenous kampungs. The reforms aimed at improving the conditions for the native population typically remained only partially implemented and some of the seemingly benign policies actually proved detrimental.

The new decentralized governance system took effect in 1905. One of the first acts by the new municipal council was to institute greater regulations over the urban development process. A 1909 law passed by the Batavia Municipal Council established the city's first codified regulations governing building standards. It placed responsibility for issuing building permits in the Municipal Works Department, whose staff of architects and engineers prepared the city plans and models for new houses. The director of the Municipal Department of Land and Housing, F.J. Kubatz, prepared Batavia's first comprehensive plan in 1918.[48] In that same year, the Batavia government created the Municipal Water Company (*Water Leidingen Bedriff van Batavia*) to manage the city's expanded system of water. The water company provided piped spring water directly to houses in the European neighborhoods and to the native communities by means of community water taps. Improved water delivery sought to improve the health conditions in Batavia neighborhoods. There were wide-ranging discussions of appropriate ways to encourage natives to stop using the canals as a water source, the favored being extending the European water system to the native communities through public pipes.[49] Between 1922 and 1926, spring water supplied to the native communities from public hydrants was free of charge. This was done with the intention to improve public health and as a response to the Dutch government's mandate to foster a modernized, productive, and efficient native population. Free water compromised the financial health of the municipal water supply company, however, owing to the high costs of constructing the piping to access spring water. Citing financial exigencies, the water company questioned the efficacy of the native population getting free water and consequently stopped that practice. They justified the termination of free water on the grounds that a new system of "valuation" around water supply would help to develop more economically efficient users. The argument was that "[w]ater that was 'paid for' would 'gain more value in the eyes of the population so that it is no longer wasted in despicable ways.'" When the water authority changed from free to "paid delivery" in 1926, it formally appointed native water vendors to sell water from the public hydrants at a predetermined profit margin. The addition of an intermediary increased the cost for the consumers beyond what they would have paid directly to the water company.[50] Another water management practice introduced during this period involved drying out the coastal areas where standing water served as a breeding place for malaria-carrying mosquitoes.

> Parallel to the coast and for a total of length of 4k, the low-lying area to the north of the city was dried out as much as possible; chiefly by means of improved direct draining during periods of ebb tide.[51]

The 1918 Kubatz plan for the expansion of the Batavia's settlement incorporated the van Breen flood canal proposal to protect the newly developing European community in the area of New Gondangdia well south of the original settlement. This included the newly planned garden city of Menteng. The original plan for Menteng, the work of Dutch architect P.A.J. Mooijen, developed when the Batavia city government sanctioned the creation of a development group, the *Commisie van toesicht op het Geheer van het Land Menteng*. This public company was responsible for planning and developing the larger area of Gondangdia (*Nieuw Gondangdia*). The Kubatz plan for the nearby flood canal modified the original Mooijen plan for Menteng. This new plan called for the flood canal to form the southern boundary of the planned community. In addition to protection afforded by the flood canal, one of the key features of homes built in this new residential area was inclusion of modern wastewater facilities. As noted by Mrazek, unlike other areas of Batavia, in Gondangdia "the pipes for collecting refuse in the settlements are planned to be at least six meters from the limits of each of the inhabited compounds."[52]

Plans for new residential areas for the European population displaced the native communities, particularly where the lands occupied by kampungs were places without recurring flood conditions. In the case of Menteng (Figures 2.8 and 2.9), the 73 hectares of elevated land designed for the community accommodated a native peasant population of 3,562 in 1890. As Marcussen noted, the early 20th-century expansion of Batavia marked the first of three distinct periods of transformation of native kampungs in Jakarta and a tradition of kampung displacement. This was not just a consequence of Menteng's development but also a result of expansion of the Chinese community outside the Glodok ghetto after the Batavia government lifted the residential restrictions for this group in 1911. This led to growing residential densities in those native kampungs that had to accommodate those displaced by expansion of the European community.[53]

Figure 2.8 The Paulus Church was one of several Dutch Reformed churches built in Menteng to serve its predominantly European population. It was designed in 1936 by Dutch architect Frans Ghijsels who is credited with many notable buildings in Batavia, including the central train station.

Source: Photo by author.

Colonial officials became increasingly concerned with the unhealthy conditions in the densely populated kampung, mostly lacking basic water resources and sanitation facilities, and plagued with poor housing conditions. After 1915, it was customary for visits by top colonial officials from the home country to visit at least

Figure 2.9 Menteng planned community for European population.
Source: Collection of the Nationaal Museum van Wereldculturen, Coll. no. TM-20000919.

one "bad ***kampong***" to get an object lesson about the pervasive slum housing and health problems.

There was growing concern about the environmental conditions that appeared to contribute to the high mortality rates among the indigenous population. The Semarang-based pharmacist, philanthropist, and local politician Henry Freek Tillema advanced understanding of health and housing conditions in the colonies. Between 1914 and 1923, Tillema compiled a six-volume study of public health throughout the colony, including conditions in Batavia. The evidence presented in this detailed study heightened concern about the conditions in the native kampungs. He was self-described as an "engineer of health and hygienist" but others referred to him a "propagandist for clean water."[54]

Tillema's work exemplified the increased interest in urban quality-of-life matters within the colonial population. Even in Europe's most progressive city, Amsterdam, there was no comprehensive effort to provide and distribute a clean water supply to the public until the 1890s.[55] In the colonies, these efforts gained greater urgency by local government under the new decentralization policy. Beginning in 1911, annual forums in the colonies discussed the implications of government decentralization. At the Eighth Decentralization Congress held in Batavia in 1918, D. de Jongh, Semarang's new mayor (Tillema had already returned to the Netherlands to live by that point), presented a report on the housing conditions in his city and the efforts to institute sanitary improvements. He noted that urban improvements often displaced existing kampungs, and that this led to crowding in other areas. He pointed to the dual challenge of upgrading kampungs and providing new housing for the more affluent residents, efforts that sometimes were in conflict. The Social Technical Society was formed in 1918 to campaign for a public housing program. This group subsequently organized the First National Housing Congress in 1922. This led to a limited liability housing company established in Batavia in 1926, a development organization that produced a small volume of new urban housing. Overall, however, the housing company had virtually no impact on housing conditions for the poorest residents. According to one Dutch observer in 1927, the housing built was like "barracks where an occupant felt as one of a number of fellow sufferers looking for the first good opportunity to find shelter in a normal kampong."[56] Other housing alternatives came up during the first Native Housing Congress, held in Batavia in 1925. One outcome from the Native Housing Congress was the initiation of a Kampung Improvement Program. The difference between the work of limited liability housing company and kampung improvement was the shift of focus from new housing construction to finding strategies to prevent further deterioration in the existing enclaves through modest infrastructure upgrades. The advantage of upgrading existing kampung housing was that it did not necessitate displacement. In addition, improvements could be done faster and affect more families than new constructions. Calls for addressing the condition in kampung housing gained political momentum during the 1920s and 1930s as the European community became increasingly aware of the health threats they posed. During the 1930s, kampung improvement initiatives took place in all of Java's largest municipalities. In 1934, the Batavia government formalized the improvement program through passage of the Kampung Improvement Bylaws.[57]

The leading planning practitioner in the Netherlands Indies, Thomas Karsten, who drafted a planning law in the 1930s covering all East Indies cities, considered the kampung problem serious and likely to get worse since the urban improvement schemes actually accelerated displacement and densification. He considered it essential for all city plans to reserve enough suitable land for kampung expansion.[58] Existing land regulations were weak and even those enacted lacked enforcement mechanisms. Only the slow growth of Batavia's population in the 1930s, not an effective program regarding kampung improvements, reduced the pressures to find suitable new spaces for native settlements.

To support effective planning in the native settlements, a comprehensive public health research project began in 1937 that systematically gathered data on the hygiene and environmental conditions in the kampung communities. The

Study Ward for Hygiene in Batavia project, as it was called, responded to the problem that far too little was known about the qualities of densely populated sections in the city. One of the researchers noted,

> our knowledge of conditions of a far off tribe in New Guinea is greater than of a great section of the population in densely populated Java, in particular in towns ... a fact which can indeed be explained but cannot be denied.[59]

The study continued for six years, concentrating on the Tanah Tinggi area of Batavia and its 25,000 inhabitants. The researchers kept track of disease and mortality but also much more. They took into consideration "care for the environment in the broadest sense," which meant also "good houses ... built on proper sites [and] houses should have a supply of reliable drinking water. The sewage, fecal matter, urine, refuse, and drainage water should be arranged so that they cause no pollution to the surroundings."[60] As the study verified, nearly one half of the sample population lived in "unimproved" kampungs where there was insufficient drainage. This caused muddy conditions during the rainy season and contributed to standing water, which served as breeding places for mosquitoes and waterborne diseases. One important finding was that those areas that had municipally constructed dwellings were far less prone to the problems of poor drainage, standing water, and flooding during the rainy season.[61] Although flooding was a routine feature of life in Batavia (see Figure 2.10), it tended to be treated as an annual inconvenience that was an unavoidable consequence of living in a flood plain.

Although the study described the municipal water service as a potentially sufficient source, it found that only 20% of the dwelling units in the study area connected to service. Although Batavia provided public hydrants, the cost of installation in the units exceeded the means of the poor. This left most to rely on the private vendors with lower quality water and even higher prices. The numerous wells investigated in the study area were deemed to be "almost all unsatisfactory from a sanitary point of view" either because of poor construction or inadequate maintenance. The research identified 43 wells and all were contaminated. Of the 6,374 dwellings in the study area, over 5,000 suffered from inadequate water quality. There were 1,326 units with private taps and the rest served by 37 common taps.[62]

In Rawah Kampung, a Tanah Tinggi neighborhood located on the eastern side of the Sentiong drainage canal, the houses were in the poorest condition. This area was a mix of Europeans, Eurasians, and Javanese of moderate prosperity who preferred the spaciousness of a rural setting despite the lack of municipal services since it was outside of Batavia's boundary. Other sections of Tanah Tinggi benefitted by the late 1930s from the growing interest in public improvements, which included new gutters to remove rainwater. Except for the few living in the newly constructed rental dwellings with indoor plumbing, the water for most of Tanah Tinggi came from public taps supplied by the municipal water system. Chinese merchants benefitted from the process, as they sold water to residents

Figure 2.10 Flooded street in Batavia, 1918–1920.
Source: Collection of the Nationaal Museum van Wereldculturen. Coll. no. TM-30035667.

that they secured from the municipal system. Many other residents in these areas continued to rely on private wells and river water, however.[63]

Mrazek suggests that the provision of clean water and sanitary facilities for the native population was a test of colonial effectiveness at a time when critics at home and within the Netherlands Indies questioned the merits of the system. The fact that the Dutch took pride in providing free water for the residents of Batavia and then cut off access to the indigenous population found its way into published critiques of the colonial regime. In a 1927 article in the nationalist journal, *Seoloeh Indonesia* (Torch of Indonesia), it used the metaphor of water running out to suggest that the failure to adequately address native water needs foreshadowed the demise of the colony itself. Like the broader promises of the Ethical Policy, the announcement of the addition of new tap water, clean and free, came with great ceremony. Yet the festivities soon subsided when access to

> the public fountains was reduced … Eventually the remaining public fountains had opening time cut to merely few hours a day. Finally, as a crown to the whole edifice, there came – a shutting down of the public fountains altogether … Instead of water on tap that cost nothing, the guilds of water bearers and water sellers appeared on the streets.[64]

There was also great attention given to a plan in 1923 to erect in Batavia a model community for the natives, Taman Sari (Fragrant Garden) composed of seven large residential buildings, each with bathing and laundry facilities. Cost considerations caused the abandonment of the plan. Nevertheless, throughout the first four decades of the 20th century, until the Japanese seized Batavia and all of the Netherlands Indies, the issues of clean water, sanitation, slum improvement, and city hygiene were central concerns of the policy agenda. The problem was the lack of clear progress on any of these concerns apart from a deeper understanding of what was wrong.[65]

On the eve of the Japanese invasion of the Netherlands Indies in 1942, the multiple water problems that had plagued Batavia in the 19th and early 20th centuries, such as flood protection and expanded provision of clean water, had been addressed, at least partially, through modernization efforts. The flood canal and floodgates appeared to reduce the frequency and intensity of flooding, and there was an expanded supply of clean water from springs, wells, and reservoir sources to keep up with the pace of urbanization, even though there was differential treatment of the indigenous and European communities. Moreover, the spread of urbanization continued to follow the high ground to the greatest extent possible as development pushed further southward. Batavia remained compact enough to enable the existing water management system to function relatively effectively, at least for some of the population. Conditions fundamentally changed in the aftermath of the Japan's withdrawal, the failed Dutch restoration, and independence. Not only was Batavia renamed Jakarta by the Indonesians, but the compact colonial city that the Dutch nurtured gave way to what was to be a sprawling megacity. The spectacular growth of the new Indonesian capital city of Jakarta placed demands on the water and flood management systems far greater than could have been contemplated, let alone properly handled, within the compact colonial city Batavia. New challenges demanded new approaches. Unfortunately, the continuity between the colonial and postcolonial institutional approaches to planning, land management, and water management generated a new array of problems and far too few solutions for the water city.

Notes

1 Roosmalen, Pauline K.M. van (2008) "For Kota and Kampong: Re-Emergence of Town Planning on a Discipline," in Wim Ravesteijn and Jan Kop, eds. *For Profit and Prosperity: The Contributions Made by Dutch Engineers to Public Works in Indonesia, 1800–2000.* Leiden: KITLV, pp. 272–302.

2 Raffles, Thomas Stamford (1817) *A History of Java.* London: Black, Parbury, and Allen, Vol. 2, Appendix viii.

3 Ibid.

4 Reid, Anthony (1988) *Southeast Asia in the Age of Commerce, 1450–1680.* New Haven: Yale University Press, pp. 212–213; *Jakarta Post,* July 10, 1999; *Jakarta Post,* May 8, 2003.

5 Abeyasekere, Susan (1989) *Jakarta: A History,* Revised Edition. New York: Oxford, p. 6.

6 Blusse, Leonard (2008) *Visible Cities: Canton, Nagasaki, and Batavia and the Coming of the Americans.* Cambridge: Harvard University Press, p. 18.
7 Ibid., pp. 23–25.
8 Abeyasekere (1989), pp. 9–13.
9 Blusse (2008), pp. 60–63.
10 Ibid., p. 65.
11 Cobban, James L. (1985) "The Ephemeral Historic District in Jakarta," *Geographical Review,* 75, no. 3 (July): 301.
12 Caljouw, Mark, Nas, Peter J.M., and Pratiwo (2004) "Flooding in Jakarta," paper presented at First International Urban Conference, August 23–25, Surabaya, p. 2.
13 Abeyasekere (1989), p. 15.
14 Heuken, Adolph S.J. (1982) *Historical Sights of Jakarta.* Jakarta: Cipta Loka Caraka, p. 28.
15 de Haan, *Oud Batavia,* 2nd Ed., Vol. 1, pp. 46 and 96, cited in Cobban (1985), p. 301.
16 Blusse (2008), p. 15.
17 Cobban (1985), p. 301.
18 Ibid., p. 303 (map by Clemendt de Jonge cited in *Oud Batavia*).
19 Cobban (1985), p. 305.
20 Blusse (2008), p. 15.
21 Heuken, op. cit., pp. 154–157.
22 Blusse (2008), p. 15.
23 Jong, Frida de and Ravesteijn, Wim (2008) "The Rise and Development of Public Works in the East Indies," in Wim Ravesteijn and Jan Kop, eds. *For Profit and Prosperity: The Contributions Made by Dutch Engineers to Public Works in Indonesia, 1800–2000.* Leiden: KITLV, pp. 51–52.
24 Blusse (2008), p. 27.
25 Caljouw, Mark, Nas, Peter J.M., and Pratiwo (2005) "Flooding in Jakarta: Towards a Blue City With Improved Water Management," *Bijdragen tot de Taal-, Land-en Volkenkunde,* 161 (4): 467.
26 Blusse (2008), pp. 27–28.
27 Ibid., pp. 29–34; Kathiritham-Wells, J. and Villiers, John, eds. (1990) *The Southeast Asian Port and Polity: Rise and Demise.* Singapore: Singapore University Press.
28 Raffles, op. cit., p. 246.
29 Abeyasekere, Susan (1987) "Death and Disease in Nineteenth Century Batavia," in Norman Owen, ed. *Death and Disease in Southeast Asian History.* Singapore: Oxford University Press, pp. 189–209.
30 Ibid., p. 194.
31 Stockdale, John J. (1811) *Sketches, Civil and Military of the Island of Java and the Immediate Dependencies: Comprising Interesting Details of Batavia, and Authentic Particulars of the Celebrated Poison-Tree.* London: J.J. Stockdale, pp. 128–133.
32 Ibid., pp. 57–60.
33 Ibid., pp. 133–134.
34 Kooy, Michelle and Bakker, Karen (2014) "(Post) Colonial Pipes: Urban Water Supply in Colonial and Contemporary Jakarta," in Colombijn, Freek and Cote, Joost, eds. *Car, Conduits and Kampongs: The Modernization of the Indonesian City, 1920–1960.* Leiden: Brill, p. 66.
35 Abeyasekere (1987), p. 196; Abeyasekere (1989), pp. 70–71.

36 Wit, Augusta de (1912), *Java: Facts and Fancies.* The Hague: W.P. van Stockum; Reprinted by Oxford University Press, Singapore, reprinted 1984, p. 29.
37 Ibid., pp. 43–47.
38 Mrazek, Rudolf (2002) *Engineers of Happy Land: Technology and Nationalism in a Colony.* Princeton, NJ: Princeton University Press, pp. 55–56.
39 Ibid., p. 56.
40 Veering, Arjan (2008) "The Formation of the Port System," in Wim Ravesteijn and Jan Kop, eds. *For Profit and Prosperity: The Contributions Made by Dutch Engineers to Public Works in Indonesia, 1800–2000.* Leiden: KITLV, pp. 192–237.
41 Ibid., pp. 204–214.
42 Abeyasekere (1989), pp. 49, 208n26.
43 Kop, Jan (2008) "Drinking Water, Sanitation and Flood Control in Urban Areas," in Wim Ravesteijn and Jan Kop, eds. *For Profit and Prosperity: The Contributions Made By Dutch Engineers to Public Works in Indonesia, 1800–2000.* Leiden: KITLV, pp. 314–319.
44 Quoted in Mrazek, op. cit., p. 57.
45 Caljouw, Nas, and Pratiwo, op. cit., p. 470. See also: Algemeene Secretarie No. 904, 4 Februari 1948 Recanavan Breen cukuk sertiana traversal channel.
46 Indonesia, National Archives (2003) *Masalah Banjir Batavia abad XIX, Proyek Pemasyarakatan dan diseminasi Kearsipan Nasional.* Jakarta: Arsip Nasional Republic Indonesia, p. 3.
47 Gouda, Frances (2008) *Dutch Culture Overseas: Colonial Practice in the Netherlands Indies, 1900–1942.* Amsterdam: Amsterdam University Press, pp. 23–24.
48 Akihary, H. (1996) *Ir.F.J.L. Ghijsels: Architect in Indonesia, 1910–1929,* translated by T. Burrett. Utrecht: Seram Press, pp. 14–15, 19–23, 94, 98.
49 Kooy, Michelle and Bakker, Karen (2008) "Splintered Networks: The Colonial and Contemporary Waters," *Geoforum,* 39 (6): 1843–1858.
50 Ibid.
51 Kop, op. cit., p. 328.
52 Mrazek, op. cit., p. 56. For a detailed discussion of the Menteng plan, see Silver, Christopher (2011) *Planning the Megacity: Jakarta in the Twentieth Century.* London: Routledge, pp. 56–60.
53 Marcussen, Lars (1990) *Third World Housing in Social and Spatial Development: The Case of Jakarta.* Aldershot: Avebury.
54 Quoted in Mrazek, op. cit., p. 57; Cote, Joost (2002) "Toward an Architecture of Association: H.F. Tillema, Semarang and the Construction of Colonial Modernity," in Nas, P.J.M., ed. *The Indonesian City Revisited.* Singapore: Institute of Southeast Asian Studies, pp. 319–347.
55 See Mrazek, op. cit. p. 246n67.
56 Cobban, James L. (1993) "Public Housing in Colonial Indonesia, 1900–1940," *Modern Asian Studies* 27 (4): 893–894; Silver, op. cit., p. 65.
57 Kampung Verbeeterings Ordonnantie (1934); Tesch, J.W. (1948) *The Hygiene Study Ward Centre at Batavia: Planning and Preliminary Results, 1937–1941.* Leiden: University of Leiden; Flieringa, G. (1930) De Zorg Voor de Volkshuisvesting in de Stadsgemeenten in Nederlandsch Oost Indie, in *Het bijzonder in Semarang.* S-Gravenhage: Nijhoff, cited in Polle, V.F.L. and Hofsee, P (1986) "Urban Kampung improvement and the use of aerial photography for data collection," in Nas, P.J.M., ed., *The Indonesian City: Studies in Urban Development and Planning.* Dordrecht: Foris Publications, p. 119; Abeyasekere (1989), pp. 120–122; Silver, op. cit. p. 65.

58 Bogarers, E. and Ruijter, P. de (1986) "Ir Thomas Karsten and Indonesian Town Planning, 1915–1940," in Nas, P.J.M., ed. *The Indonesian City: Studies in Urban Development and Planning*. Dordrecht: Foris Publications; Wertheim, Wim F. (1958) *The Indonesian Town: Studies in Urban Sociology*. The Hague and Bandung: W. van Hoeve, p. 68.
59 Tesch, p. 26; Silver, p. 70.
60 Tesch, p. 17; Silver, p. 71.
61 Tesch, pp. 199–201.
62 Ibid., pp. 101–104.
63 Tesch, p. 53; Silver, p. 72.
64 Quoted in Mrazek, p. 59.
65 Mrazek, p. 60.

3 Water management in the new capital

Independence from the Dutch and the beginning of the Indonesian republic, effectively underway in 1950, transformed the colonial administrative center of Batavia into the new (but still unofficial) capital city and the visible hope of an aspiring new nation. The creation of a visionary plann for the capital in the 1950s and 1960s was essential to the national identity-building process. But the day-to-day concerns focused on several major objectives, one being to provide housing and services to accommodate its rapidly growing population. The other was to modernize the city, an obsession of President Sukarno, who tended to put the ceremonial and grandiose ahead of the more mundane needs of an inadequately resourced big city with substantial unmet basic services. Plans to expand Jakarta deep into the agricultural areas to the south of the city had already been in the works by the Dutch authorities. They had expected to resume control of their East Indies empire following expulsion of the Japanese and despite the 1945 Indonesian proclamation of independence. As the battle to regain their colony over the next four years, the Dutch government set up a Central Planning Bureau under the Department of Public Works and Traffic to plan for postwar reconstruction. Jacques P. Thijsse, a civil engineer who had worked in the Netherlands Indies since 1921, served as its director.[1]

One postwar project the planners intended to usher in environmental upgrading for the restored colonial center was a regional flood control project formulated by Dutch hydrologist Willem Johan van Blommestein and released in 1948. His plan dealt with irrigation and drainage throughout West Java and included recommendations for constructing dams and lakes to capture the waters of the Citarum to increase utilization of surface water supplies. This included a plan to construct hydroelectric power facilities and to expand irrigated agricultural plots. For flood protection, van Blommestein proposed the construction of a dike stretching eastward from Batavia to the Tanjung Priok. The dike would create a system of polders (following the standard practice to address flooding in the Dutch system), but Blommestein also pointed to the need to dredge the canals and rivers, to add new pumping stations and sluices, and to expand distribution of drinking water. Obviously, the failure of the Dutch to restore hegemony over the Indonesian colonies, and the lack of interest at the time by the new Indonesian republic government to take up Blommestein projects, meant that none of the

DOI: 10.4324/9781003171324-4

suggested actions were undertaken. Sixty years later, and in response to recurring devastating floods, a scheme similar to the coastal dike that Blommestein proposed would become the centerpiece of a bold but controversial flood control scheme tendered by another group of Dutch consultants (and discussed in Chapter 5).[2]

Rather than flood control or expanded water service, Jakarta's postwar reconstruction centered on expanding housing for the Indonesians now in charge of the government. Development of a new satellite city, Kebayoran Baru, built on a 750-hectare tract that formerly functioned as fruit groves, was the most ambitious one of the housing projects. This was another reconstruction scheme prepared by Dutch experts but this one fitting with the agenda of the Indonesian government. The original scheme was prepared by a faculty member at the Institute of Technology in Bandung, Professor Ir. V.R. van Romondt. The site selected for the planned satellite city was high ground five kilometers south of the current Jakarta boundary on land originally intended for a new airport. The two bordering rivers, the Kali Krukut to the east and the Kali Grogol to the west, and the undeveloped lands along these waterways were to serve as a greenbelt surrounding Kebayoran Baru. Together, they created a natural growth boundary for the settlement. The plan acknowledged that these adjacent low-lying lands routinely flooded and thus should remain as open space. Implementation of the Kebayoran Baru plan by the Indonesians after independence stayed true to the original settlement scheme design except that no effort was made to preserve the greenbelt. The pressure to build for the rapidly expanding population, no matter if the land was in a flood zone, trumped the need to regulate growth. Proximity to the fashionable Kebayoran Baru community was an attractive calling card for new housing. It did not take long for housing appearing in the low-lying areas to be routinely flooded during the rainy season.[3] When massive floods began to cripple parts of the city in the mid-1990s, it was these areas that were inundated in the otherwise unaffected sections of South Jakarta.

Groundbreaking for Kebayoran Baru occurred in 1949, months before the final transfer of full authority from the Dutch to the Indonesians, with Thijsse remaining after the European exodus to guide its implementation. The construction of the community was under a special Kebayoran Baru Authority (KBA) headed by an Indonesian engineer, Srigati Santoso. The KBA was responsible for site work, constructing all of 120 kilometers of street network within the satellite city, and completing the 4.5-kilometer highway to connect Kebayoran Baru to Jakarta proper. In 1950, Jakarta annexed Kebayoran Baru along with a vast swath of undeveloped peripheral lands, more than doubling the land area of the city. Over the next several decades, this larger Jakarta experienced intense development that expanded the city's footprint well beyond the compact but now severely overcrowded colonial capital.

The Kebayoran Baru plan called for the construction of over 7,000 housing units to address the housing shortage, including a mix of houses for the middle class families, especially those working in civil service, as well as 1,713 smaller low-cost units. The actual number of units built reached 4,720, and none proved accessible to Jakarta's burgeoning poor population. The size of houses originally slated for the poor (called *rumah rakyat*, or people's house) was 68 square meters and included three bedrooms, a living room, kitchen, and interior bathroom. All of the houses, including the *rumah rakyat*, were to connect to the expanded clean water system that was part of President Sukarno's larger modernization efforts. Although Kebayoran Baru did not have an enclosed sanitary sewer system, the community got 12,302 meters of drainage channels for wastewater and another 50,283 meters of channel to handle rainwater. The original plan called for the central water system to link to a water purification plant but at least initially the houses were supplied with water from artesian wells located within the community.[4]

Concurrent with the development of Kebayoran Baru, the Sukarno government was financing two surface water treatment plans, *Pejompongan* I and *Pejompongan* II, to expand the supply of water (and to compensate for abandoning the use of spring water) by making greater use of surface sources. *Pejompongan I* began operation by 1957 and added 2000 l/s to the water supply and by the mid-1960s, *Pejompongan II*, added another 1000 l/s. This supported many of the new modern spaces constructed to help realize Sukarno's quest to distinguish the new capital city from its colonial forbearer. Yet as in the colonial era, as Kooy and Bakker emphasize, the expanded postcolonial urban water supply had a narrow reach. It remained limited to modern areas of the city much like the colonial system served predominantly the European community.[5]

Between 1940 and 1960, the average population density of Jakarta dropped because of the land area added in the 1950 annexation. At the same time, the massive in-migration that swelled the city's population during the 1950s and continued to grow rapidly through the 1980s quickly offset the initial density drop. More than any other single factor, the scale and rapidity of population in-migration contributed to the loss of control by the government of the urbanization process and overwhelmed its already inadequate services, including the water management system. From 1950 through the early 1970s, the population of Jakarta grew from roughly one million to more than four and one-half million. Estimates had suggested that the population would not reach that mark until the end of the 1970s. In fact, the city reached the four million mark by 1971, as a result of average annual population growth in the 1950s of approximately 133,000 that increased in the 1960s to approximately 167,000 annually.[6] Although the newcomers scattered throughout the metropolis, the city's planners attempted to direct new urban housing developments to accommodate the growth around three regional centers; Tangerang to the west, Bekasi to the east, and Bogor to the south.

Owing to the shortage of housing to accommodate this substantial population increase, the existing kampungs were crowded with new inhabitants, and much of the vacant land along the city's 13 rivers, which afforded free access to water, proved attractive places to construct makeshift settlements. Previously sparsely settled riverbanks and canal-front lands quickly became home to thousands of migrants. There were no controls in place to prevent the occupancy of these vacant and marginal spaces, since the planning function for Jakarta dealt largely with envisioning the larger sweep of development rather than focusing on the daily incremental transformations that accompanied the dynamic change processes underway.

Planning for the city as a whole, as distinguished from community-level plans such as the Kebayoran Baru satellite city, got underway in the 1950s with assistance from the United Nations. A primary objective of this planning process was to contrive a blueprint for the spatial ordering of the capital city and its environs. The first comprehensive plan, commonly referred to as the Outline Plan, was, in fact, just an outline. Completed in 1957 and officially endorsed in 1959 by the Municipality of Greater Jakarta "as a basis for the future development of the city," it provided a basic framework for planning but not specific initiatives to realize its vision on the ground.[7] According to the recommendations of the consultant team that helped to prepare it, headed by British planner Kenneth Watts, it was government's responsibility to handle the "social overhead," namely to address the need for a universal water supply, upgrade the drainage systems, and construct needed roads, with the private sector handling the city's dire need for additional housing. Despite these recommendations for broad-based public infrastructure investments, in the areas of water and sewerage especially, the policy stance adopted by the Indonesian government ran contrary to the plan's recommendation. According to the Jakarta government, water and sewerage were not public responsibilities but rather a "private concern." This stance would have profound implications for a water management system that had to cope with ever-increasing demands for service by a population too poor to bear the costs on their own.[8]

The Outline Plan provided the basis for a master plan prepared in 1960 that conceptualized Jakarta's future development in terms of concentric rings emanating from the core in what had been the initial settlement (*kota*) and the 19th-century extension (Weltevreden). Unfortunately, the conception of Jakarta's future development as three distinctive concentric zones did not mesh with the historical pattern of settlement dictated by the geography where the city had been placed. Jakarta's growth since the early 19th century had been linear, following the highest possible ground between the 13 rivers that flowed northward from the mountains into the Bay of Jakarta. Within this linear development configuration, Jakarta had acquired several distinct and separate commercial centers while at the same time accommodating population growth through high densities on sites less prone to flooding. One digression from the traditional north-to-south development pattern was the suggestion to create additional satellite cities like Kebayoran Baru but emanating in an arc from the west and to the east along the city's outer boundary. The idea was to better concentrate development away from center while simultaneously addressing the housing and infrastructure deficits apparent in the overcrowded older sections of the metropolis.[9] But this also meant pushing development from the high spine onto the lower lands, often marshy areas, that crossed some of the 13 rivers. One of these satellite cities was Pulo Mas, originally planned as a full-service community to accommodate a population of 50,000 workers by means of a series of interconnected neighborhood units. Besides new housing, the Pulo Mas plan called for state-of-the-art infrastructure, including clean water taps to support residents and then later piped water in individual dwellings. The Pulo Mas master plan also called for a system of closed sanitary sewers connected to each dwelling and tied into a sewage treatment scheme that would ensure primary treatment before discharge into the nearby Sunter River. A system of surface drains would handle rainwater, taking the runoff through open ditches to a newly constructed reservoir to prevent flooding especially during the rainy season. Since this marshy area previously had served as rice fields, it was an area prone to flooding.[10]

Construction of Pulo Mas began in the mid-1960s under the governorship of Ali Sadikin. While some of the infrastructure upgrades occurred, the original intent to accommodate low-cost housing changed to lower density upper-income units. The city scrapped the original plan to draw upon public funds to develop the community because of the dire state of Jakarta's finances in 1966. Sadikin's alternative approach was to use public power to secure the land from private owners and allow private developers to finance construction. The mass housing scheme initially conceived was not financially feasible for the private sector without a public subsidy, and this was not available for low-income housing in the 1960s. What was feasible was more upscale housing to serve the middle-income market at significantly lower densities. Apart from a small housing complex north of the community, known as Rumah Susun Pulomas, development of Pulo Mas did nothing to alleviate the housing pressures faced by working-class families in the rapidly growing city. In fact, a dense concentration of squatter housing along the Sunter River emerged adjacent to the new housing of Pulo Mas.[11] Following the development of Pulo Mas, a robust satellite city-building effort took place in the 1980s and 1990s, a process dictated by private developers supplying middle- and upper-income housing and often uprooting existing lower-income communities to accommodate them.

The 1965–1985 Master Plan of Jakarta, adopted in May 1967, captured through its analysis the emergence of the ring of suburban communities that was consistent with the Outline Plan and that would become the locus of urban expansion over the next several decades, as shown in Figure 3.1. The plan also identified five major problems that needed resolution as the metropolitan area developed over the next 20 years. Virtually all of these involved basic services and infrastructure. One was the flooding that became more problematic as new settlements appeared in flood-prone areas. In response to this, the central government initiated the "Komando Proyek Pencegahan Banjir DKI Jakarta" (Command of Jakarta Flood Prevention Project). To implement this initiative, Jakarta established an agency to address flooding in Jakarta, with jurisdiction over reservoir/basin construction, river rehabilitation, polder construction (five polders), and river diversion. Improved management of sanitation, especially the removal of solid waste, was another problem that called for an improved overall system of collection and disposal.[12] Traffic congestion was another matter

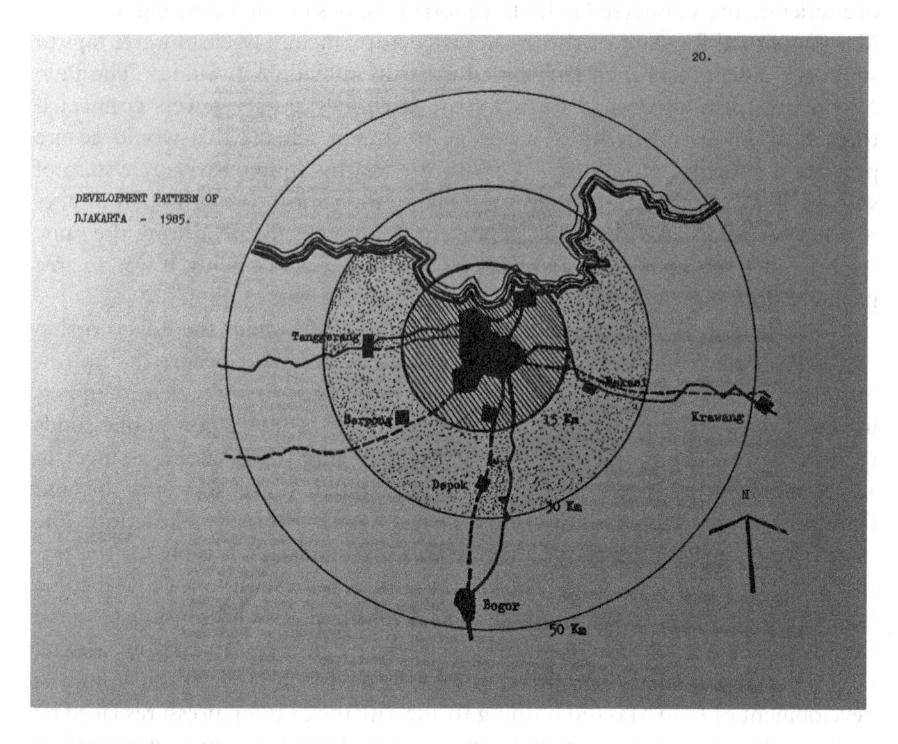

Figure 3.1 Sketch from the 1965–1985 master plan showing the scheme for development of the region toward the key satellite cities.

Source: DKI Jakarta Master Plan, 1965–1985.

demanding immediate intervention. As the 1965 plan noted, the transport problems emanated from a combination of mixed land uses, poorly designed road intersections, the existence of surface rail lines crossing major thoroughfares, and the recurring flood conditions.[13]

The greatest challenge confronting the city's planners was accommodating the rapidly growing population streaming into Jakarta from the countryside. Jellinek's (1991) classic study of community life in the Kebun Kacang kampung in Central Jakarta from the 1930s through the early 1970s shows how one place adapted to these pressures. Early arrivers in the 1930s chose higher ground to the west of the Cident River, but those who came *en masse* in the 1950s and 1960s occupied land next to the river that routinely flooded during the "rainy season." As Jellinek put it:

> The river swelled with each monsoon season and almost always overran its banks, sometimes repeatedly. The turgid, pungent waters, often up to a metre in depth, took over the kampong … Families moved themselves and their few possessions onto their roofs or into the houses of friends if they

> were lucky enough to escape the deluge … They waited for the waters to subside and then set about re-establishing their lives. But the smell of the receding waters, the watermarks on the walls and the outbreaks of gastro-enteritis or other water-borne diseases had first to be patiently endured. After the floods, the earthen pathways turned to a quagmire so that one had to wade, knee-deep in mud.[14]

An updated version of Kampung Improvement Program (KIP), originally conceived by the Dutch reformers in the 1920s (as discussed in Chapter 2), became the signature effort under the Sadikin's governorship to improve environmental conditions in poor kampungs throughout the city. Sadikin's KIP paid particular attention to the construction of new elevated hard surface walkways to eliminate the muddy pathways during the floods. Yet as Jellinek discovered in Kebun Kacang, these walkways might carry pedestrians above the floodwaters, but also they caused the water to flow into the residences that were now like buckets collecting the runoff from the paved paths.[15] Adding concrete walls to the rivers, such as in the case of the Krukut River as it passed through South Jakarta (see Figure 3.2) afforded additional protect while at the same time eliminating the green areas adjacent to the rivers that could absorb high waters.

Figure 3.2 Section of the Krukut River passing through built-up area of Central Jakarta. Source: Photo by author.

Kampung Kebun Kacang occupied low-lying land adjacent to the expanding commercial center along Jalan Thamrin. As pressures increased to find new sites for commercial uses in Central Jakarta, the marshy, muddy lands that made up much of Kebun Kacang suddenly became prime land for redevelopment. Whereas flooding conditions had not driven the poor from the area, commercial and residential redevelopment eventually did. Kampung Kacang residents first learned of the plan to raze structures and redevelop the community in the early 1980s. Although the government had invested funds in kampung improvement in this community, the KIP was regarded by city leaders as a temporary measure. The land on which the community sat was designated in the city's plan for commercial development along nearby Jalan Thamrin. So it was not pure chance that the city's first luxury international shopping mall opened in 1989 on the edge of the neighborhood that previously accommodated kampung housing. Condos and high-rise office towers followed in the 1990s on raised ground to prevent future flooding.[16]

Jakarta's kampungs huddled next to, and in some cases over, the Jakarta's rivers and canals, as shown in Figures 3.3 and 3.4. In Kali Anyar, another low-income community in the northwest section of central Jakarta and situated along the Banjir (Flood) Canal built by the Dutch during the 1920s, flooding

Figure 3.3 Kampung housing extending over the Krukut River in Central Jakarta.
Source: Photo by author.

Figure 3.4 Street drainage into the river by design.
Source: Photo by author.

was routine. In a study of health and crowding conditions conducted there under the United Nations Habitat auspices, those afflicted by flooding seemed a smaller portion of the community than in Kebun Kacang. Of the 559 household structures included in this study sample, 110 housing units reported regular flooding in the front of the house while just 13 had problems with water entering the structure. The larger problem for Kali Anyar was insufficient drinking water. Only 30% of the community had piped public water and even this was in shared facilities rather through private indoor plumbing.[17] In the case of Kali Anyar, there was no development pressure threatening the community but also there was little impetus to address its deficiencies because of it was not within an area desired by developers. So the lack of service and the lack of attention was what helped to preserve it.

The frequency of flooding in kampungs such as Kebun Kacang and Kali Anyar occurred because of poorly maintained drainage facilities, rivers filled with debris carried down from the mountains, and the lack of solid waste service. The waterways also served as receptacles for refuse and waste not collected in the kampung. The result of all these items finding their way into the rivers was a reduced capacity to handle the heavy waters during the rainy season rains. Although the mass construction phase in the periphery of Jakarta was still to come, already critical open spaces needed to absorb the rain and reduce runoff had disappeared. In response to this problem, the Office of Public Works initiated a program of reservoir construction, some limited dredging of rivers and estuaries, the addition of pumping capacity to remove floodwaters, and the addition of new floodgates to manage the water flows. The work began under the Sadikin governorship between 1965 and 1967 as the transfer of national authority from the Sukarno administration to the Suharto New Order government took place. The full scope of Sadikin's water management upgrades remained incomplete owing to ongoing budget constraints facing the Jakarta government. As a result, less than 11% of the housing sites in neighborhoods in the city had piped water. According to a local authority, "sewage and garbage disposal are already so acute that they threaten the health of thousands if not millions of the city's inhabitants."[18]

A regional approach to water management

In response to the demands for housing for Jakarta's rapidly growing population, developers sought sites in less expensive lands in the periphery. Beyond the boundaries of Jakarta, there was no plan guiding this dispersed development or any consideration of how it might impact the region's hydrology. Donor agencies, including the World Bank and the Dutch national government, advocated development of a regional planning framework to guide Jakarta's outward expansion. The Jabotabek Metropolitan Development Plan process took place from 1973 through to the early 1990s. It advocated an integrated development approach to link land development strategies to needed infrastructure within the city and its expanding periphery. The term "Jabotabek" (the acronym formed by merging the first letters of **Ja**karta, **Bo**gor, **Ta**ngerang, and **Bek**asi) recognized the involvement of four separate governmental bodies that had a stake in the development of the regional area. Tangerang and Bekasi were satellite enclaves that later developed into cities ranging between three and five million inhabitants. Later, the village of Depok, located on the southern border of Jakarta, emerged as another rapidly growing independent city that required changing the regional nomenclature from Jabotabek to Jabota<u>de</u>bek (the added "de" for Depok) by the late 1990s. Fundamental to the Jabotabek planning effort was the identification of the desired land development strategy to ensure that the location of new infrastructure would properly shape and serve the urban development needs. The Dutch consultants incorporated into the Jabotabek planning framework a strategy applied to the peripheral development in the Netherlands, namely "bundled deconcentration." It recognized the inevitable spread of development into the periphery but with a recognition that concentrating this growth would support improved transportation. The Netherlands accomplished this by various transit options. In Jakarta, the deconcentration strategy was automobile based. This pattern of guided development was supported by an east-west highway and intersected by a north-south leg extending from eastern Jakarta south to Bogor. According to the consultants, the bundled deconcentration strategy "held the promise of serving to safeguard the open spaces" in-between the urban clusters in order to safeguard lands needed for refreshing Jakarta's groundwater supply. Although the development nodes of Bekasi and Tangerang intensified as planned, overdevelopment on a more dispersed pattern in the southern region, especially in the Depok area, "accelerated the loss of undeveloped open space need for replenishment of groundwater supplies."[19] So by the early 1990s, the plan was changed to focus exclusively on development along the northern corridor to protect the southern slopes which were sources of the rivers that flowed into Jakarta as well as major water catchment area.

In the case of water management as well, a regional approach made far more sense since the rivers that passed through Jakarta and that were the source of the city's water supply, as well as the source of flooding and pollution, were regional resources. Under the Minister of Public Works Radinal Moochtar, a bureau for planning and human settlements was set up to support the regional planning effort. The Jabotabek Regional Plan articulated the concept of preservation of green spaces by channeling new urban growth into a system of new towns strategically located adjacent to protect water recharge sites. The idea was to control "wild kampong development" by forcing growth into concentrated clusters, linked together and to the urban core by improved transport. The bundled concentrations would follow a linear system, extending along highways eastward toward Bekasi, westward toward Tangerang, and southward toward Bogor.

An important feature of the Jabotabek plan was preservation of landscaped buffers, especially between Bogor and Cibinong, to safeguard aquifer recharge areas that replenished the groundwater accessed by public and private wells throughout the region. This was different from the greenbelt proposed in the Jakarta Outline Plan of the 1950s. The Jabotabek strategy discouraged development in those places that the plan identified as crucial to sustaining groundwater resources, and where runoff from projects added to river pollution. Other areas to be insulated from new urban development were the farming areas north of Bekasi and north of Tangerang, both adjacent to the sea. These were recommended to be "permanently reserved for rice production," with the caveat that if after 1985 they were needed to absorb spillover urban development, then the uses there could be changed. In the case of the area north of Tangerang, there was already a separate plan to utilize a section of the rice and fish farming area to construct a new international airport. Indeed, flooding on the access road to the Soekarno-Hatta International Airport, constructed in the late 1980s, routinely shut down the facility over the next two decades. Neither the Jabotabek Plan, nor the annual episodes of flooding in these lowlands, inclined Jakarta's leadership to prevent the relentless march of urbanization across these environmentally sensitive lands.

Development and the wishes of powerful developers, not the planners, controlled the urbanization process.[20]

Revision of the Jabotabek regional plan took place in the early 1980s, with an eye toward synchronizing with a revised master plan underway by the Jakarta government. The 1980s Jakarta plan intended to revise substantially the guidance provided in the 1965 master plan. As in the 1965–1985 master plan, the new Jakarta Structure Plan 1985–2005 focused largely on land management issues in order to set forth a process of orderly expansion given the high rate of population growth. To reduce the land needed for highway construction, it proposed improvements to public transportation and recommended limiting car usage. These seemed to be empty promises since the plan contained no real implementation mechanism and seemed contrary to scale of public and private investments underway. The plan also proposed a series of environmental initiatives to improve river water quality, to channel new urban development toward the east and west through the construction of toll roads, and (as in the Jabotabek plan) to halt growth southwards into prime agricultural areas. Here again, there was no clear implementation strategy tied to these policies.[21]

Also coming out the renewed relationship between Indonesia and the Netherlands was detailed technical advice on flood risk management. In 1973, Indonesia's Ministry of Public Works, with technical support from the Dutch firm Netherlands Engineering Consultants (NEDECO), prepared a Master Plan for Drainage and Flood Control of Jakarta. Although the drainage and flood control plan focused exclusively on DKI Jakarta, it treated Jakarta's problems as regional in scope that warranted a regional approach. The new Dutch plan proposed an extension of the van Breen Western Floodway constructed in 1924 to add the Grogol, Sekretaris, and Angke rivers to the Ciliwung Cident and Krukut that were already connected to the original canal. An Eastern Floodway was proposed to divert water from five other rivers (Cipinang, Sunter, Buaran, Jatikramat, and Cakung). Together, these flood canals could handle 100-year floodwater levels, calculated as a volume of between 290 and 525 cubic meters per second. The existing old river channels were calculated as capable of handling at best the water volume of a 25-year flood. A major flood occurred in 1979 before any of the proposed improvements had been made, thus lending a sense of urgency to the floodway construction. Given the high cost of flood canal construction and the need for a quicker remedy, the government scrapped the plan to extend the Western Floodway in favor of constructing the Cengkareng Drain, designed to withstand 100-year floods. The Cengkareng Drain was constructed between 1981 and 1982 and channeled waters from the Gropol, Sekretaris, and Angke rivers. Because of the land acquisition difficulties, the larger Eastern Floodway project remained on the drawing boards. The other component of the 1973 master plan was to improve the functioning of reservoirs and pumping stations, which proceeded slowly but steadily.[22] A second new floodway, the Cakung Drain, was built to divert runoff from Sunter, Buaran, Cakung, and Jatikramat rivers to help protect the eastern side of Central Jakarta.

In 1974, water control and management was made the responsibility of the central government under the Ministry of Public Works and empowered to cover flood control management under this function (Law No. 11/1974). It was not until 1982 that implementation regulations were promulgated (Government Regulation 22/1982) to activate its flood management functions. In the meantime, a severe flood occurred in Jakarta in 1979 that had it greatest impact in the western fringe communities. As a result the idea of an eastern canal set forth in the 1973 master plan was abandoned as a priority. This was the decision of the Directorate General of Water Resource Development in the Ministry of Public Works, a decision that would be revisited and debated numerous times over the next three decades.[23]

In 1978, the Indonesian government acknowledged the importance of environmental concerns when it created the post of State Minister for Population and the Environment. Emil Salim, a distinguished economist and avid environmental activist, was appointed as minister by President Suharto in 1983 and served in that post for ten years. During his term as minister, an environmental enforcement agency, Badan Pengendalian Dampak Lingkungan (BAPEDAL), was created in 1991 to address point-source pollution problems. One of Minister Salim's goals was to strengthen public awareness of environmental concerns, and he personally led tours down the Ciliwung to publicize the importance of river cleanup efforts. In one sense, this was the beginning of government awareness of the need to engage Jakarta's public in solutions to water problems. But as evidenced in the Ciliwung diversion project in Tangerang (discussed below), citizen engagement was not something done at the plan development and implementation stages.

In 1991, with assistance from Japan, the 1973 flood management plan was updated and expanded to incorporate concerns with wastewater disposal. The intent of the 1991 plan was to identify the requirements for a drainage system that would enable rapidly expanding Jakarta to handle its wastewater disposal needs up to 2010. But given the challenges of land acquisition, some key recommendations of the 1973 plan were not pursued, most notably the Eastern Floodway. The price tag for all the recommendations in the 1973 flood master plan was approximately 492 million rupiah, while the annual budget to handle flood mitigation in Jakarta was just 1.5 million rupiah, far short of the revenue necessary to handle annual maintenance let alone finance major infrastructure projects.

At the same time, the waterways, the riverbanks, and the adjacent lowlands continued to be filled with illegal settlements, and this contributed to the problem of keeping waterways clear. With little progress on further flood control construction, but in full recognition that Jakarta faced a growing problem of managing water, a new study of flood control in Jakarta was financed through the Indonesian Ministry of Public Works in 1996. By the mid-1990s, urbanization throughout the region meant that not just the City of Jakarta but also the surrounding satellite cities of Bogor, Tangerang, and Bekasi needed to be brought into consideration regarding flood control and overall water management. One additional new variable was a plan to reconstruct and redevelop Jakarta's north shore waterfront through a massive land reclamation effort. One of the claims of the government sponsors of the reclamation project was that it would prevent the sort of flooding that had inundated Jakarta in 1996.[24] Based upon that experience, a key tenet of the new flood control strategy was that the system needed to "reduce the load" on the Ciliwung, since that river ran through Central Jakarta and its rising waters posed a great threat to government and business centers as well as several affluent inner city neighborhoods. The plan proposed the construction of dikes and river improvements along all of the rivers in the newly urbanized areas in the periphery. It also suggested that the previously proposed Eastern Floodway be redesigned as a narrower concrete structure with high walls to reduce the amount of space needed (and hence reduce land acquisition which represented both an economic and a political problem) while maintaining the same volume of water flow.[25] Interestingly, there was no mention of regulating the location of urban settlements as part of the flood control system. Nor did the Dutch water consultants incorporate into their recommendations how those in the affected communities felt about what should be done.

It was not commonplace in the planning processes undertaken prior to the 1990s to consider the value of community participation in plan formulation. Plans typically encompassed technical and financial assessments, even though the implementation (or lack of implementation) had significant social impacts. This was especially the case in deciding the lines to follow for floodways or making improvements to existing river ways. The implementation processes clearly affected people who typically played no part in the plan formulation and who were likely to derive little or no benefits, and likely suffer as a consequence of, implementation. As Soenarno and Sasongko note, the load-reduction strategy was aimed at diverting the Ciliwung to the Cisadane. This, however, would carry the heavier discharge through the town of Tangerang. When the plan was formulated by the consultants in 1996, there were no objections raised, largely because there was no effort to engage the Tangerang community in the planning process. Yet when the construction of the project was announced in 1999, the community along the Cisadane at Tangerang, supported by its local parliament and nongovernmental organizations (NGOs), demanded to know why it had to accept the greater threat of an additional flood discharge from the Ciliwung when it already suffered from annual flooding from the Cisadane. The protest proved effective and the construction was postponed indefinitely, thereby undermining the 1996 plan.[26]

Privatizing the water supply

The problem of managing Jakarta's water supply, both groundwater and surface water, remained a critical issue throughout the 1990s. Although the government had established a regulatory permit and quota system to oversee private extractions of groundwater to protect the aquifer, since the 1960s commercial and residential use greatly exceeded the capacity of the aquifer to be recharged. Surface water sources were subject to the continuous intrusion of water draining from the streets, as shown in Figure 3.5, since there the city possessed no storm sewer capacity. Pollution of previously used surface sources placed additional burdens of groundwater sources. Groundwater levels dropped to a point, especially in northern sections of the city, where brackish water intruded into the freshwater aquifer, land subsidence was occurring, and pollution from close proximity to waste materials from adjacent commercial and industrial activities degraded water

Figure 3.5 Canal in North Jakarta passing alongside Luar Batang community which regularly experiences flooding and where one section of a seawall was built.

Source: Photo by author.

quality. In a 1991 seminar in Jakarta titled "Water, Environmental Topic Number One," J. Hillig from the UNESCO regional office for science and technology pointed to the negative effects resulting from the 3,000 registered city wells extending 100 feet into the earth. He contended that "the city faces problems related to the depletion of its water resources," with seawater intrusion occurring near the coast because of excessive pumping, and with brackish water turning up even in parts of Central Jakarta. This was compounded by the lowering of the water table by nearly three meters since the 1970s. As a result, the city has faced periodic water shortages especially during drought cycles.[27]

At the same time, the rivers and canals continued to serve as disposal points for manufacturing and domestic waste, thereby reducing their capacity to supply potable water. These conditions contributed, as Jakarta Governor Wiyogo Atmodarminto noted in 1991, to restricting the water flow during the rainy season and adding to pollution and stagnation during the dry season. Even with vigorous efforts to clean up Jakarta's surface water during the 1990s, nearly one million household septic tanks daily contributed to sewage flowing into the rivers and canals. One of the relatively clean surface water sources was the Lake Jatiluhur reservoir to the southeast of Jakarta. It functioned as a source of water for agricultural irrigation, flushing canals, and also as a raw water source for Jakarta through the Buran water purification plant.[28]

The government-owned water company, Purusahaan Air Minum Jaya (PAM Jaya) distributed water to approximately 40% of Jakarta's population, although only slightly more than 20% had piped water in their homes. The rest secured their water through community standpipes. PAM Jaya obtained its raw water from three surface sources – the West Banjir (Flood) Canal (5,600 liters per second), the Sunter River (4,000 liters per second), and the West Tarum (Flood) Canal (2,000 liters per second) – as well as several deep wells, spring water, and smaller volumes of water from other rivers and canals. Because of the polluted condition of all of these sources, the raw water PAM Jaya relied upon had to be heavily treated. Even so, it still failed to achieve World Health Organization drinking water standards. PAM Jaya also routinely suffered from insufficient supply during the dry season because during the heavy rainfall season not enough can be stored in the reservoirs. Too much of the vast amount of rainfall in Jakarta ran unused into the sea.[29] Yet like government water companies throughout Indonesia, PAM Jaya suffered not only from irregular supplies of low-quality water but also from significant water losses owing to broken infrastructure, pilfering by non-payers, and poor management.

Economist Richard Porter's study of water and waste in Jakarta in the 1990s identified eight reasons why it was prudent for the city to improve its water management system. For example, improving the quality of river water would reduce the cost of treatment. But this could only be accomplished by reducing the number of riverside residents who pollute the waterways by using them for washing, bathing, and disposing of waste. By removing residential and commercial direct inputs into the rivers, it would likely reduce waterborne diseases and thereby reduce health costs. Clean rivers might ultimately lead to an increase in the supply of edible fish. Clean rivers would also contribute to improved flood control and at the same time increase irrigation possibilities that could boost agricultural output. Finally, clean rivers would increase recreational use of the river, not to mention provide aesthetic improvements to city life. Porter contended that the general backwardness of urban environmental services in Jakarta was ironically a product of a natural abundance of water, both for drinking and for waste disposal. Signage constructed into the infrastructure politely asking residents to refrain from using the water ways for their trash, as shown in Figure 3.6, could not compensate for inadequate service levels. Yet, in the 1990s, as Jakarta's rivers and groundwater had become heavily polluted, it became "clearly necessary to undertake the investments needed by a large, crowded, and modern city to provide drinking water and waste disposal services."[30]

The water distribution system was so inadequate in 1990 that the Jakarta government passed legislation to enable those served by the water company to resell their water to neighbors who lacked access to the system. The problem with this approach was that the prices charged for water by vendors, those in charge of the community standpipes as well as the household "entrepreneurs," was anywhere from 10 to 15 times higher than the price charged to the people connected through the water company.[31] The Jakarta Water Supply Development Project called for in Jakarta's 1985 Master Plan, and funded by the Japanese development

Figure 3.6 Sign built into the wall of a Jakarta river saying "Please Don't Throw Trash."
Source: Photo by author.

Table 3.1 Water situation in Jakarta Raya, 2001

Characteristics	
Total population in service area (million)	11.0
Water production per day (cubic meters)	1,320,325
Number of utility connections, 2001	567,398
Population covered before privatization (%)	38–42
Population covered after privatization (%)	43–61
Cost of 10 cubic meters of water, 2001 (US$)	1.0
Nonrevenue water before privatization (%)	53–57
Nonrevenue water after privatization (%)	47–49
New water connections in poor areas, 2001 (%)	55

Source: Argo and Laquian (p. 231).

agency beginning in 1986 (and completed in 1997), sought to increase the water access among Jakarta households from 43% to around 82% and to reduce the unaccounted for water loss to 31%. In fact, as Table 3.1 shows, the coverage fell considerably short of that mark, and water losses continued to hover in the 47%–49% range.[32]

Table 3.2 Water sources in Kali Anyar Kampung

	Low Income (%)	*Medium Income (%)*	*High Income (%)*
Tap	12	29	59
Pumped	0	1	1
Well			
Public	51	32	19
Standpipe			
Vendor	22	21	17
Other	15	17	4

Source: Clauson-Kaas et al. (1997).

For so many, the availability of water was a costly daily exercise. A study prepared for the United Nations Center for Human Settlements of Kali Anyar, a mixed income community of 25,000 situated in the northern section of Central Jakarta, documented how its residents secured their drinking and household water (Table 3.2).

To improve water management, Jakarta's leadership hastened to join the global movement to bring the private sector into the water sector. The assumption was that by bringing private providers into the water supply system, Jakarta could salvage its failing system. In essence, this involved transforming a government monopoly into a private sector monopoly, a move made possible in the water sector by a 1994 presidential decree. President Suharto ordered the Ministry of Public Works to create a Public–Private Partnership to facilitate privatization of water service in Jakarta.[33] The privatization of water supply scheme for Jakarta involved the creation of two new firms, Kekarpola Airindo, to serve the east side of the Ciliwung River, and Garuda Dipta Semesta (GDS) to supply the western half. This unique arrangement was prompted by the desire of Indonesian President Suharto to accommodate several powerful interests. Kekarpola Airindo was a subsidiary of the powerful Salim investment group, while GDS was run by Suharto's son Sigit Hardjojudanto. Both were awarded on a noncompetitive basis. The formal water management operations for these two concerns were in the hands of the two leading international water companies, Thames Water International out of Britain and Lyonnaise des Eaux based in France. This arrangement was imposed over multiple objections from the local government water company, and 25-year contracts were signed in June 1997 and pushed to activation by President Suharto. The two private companies started operating in February 1998, right in the midst of a financial crisis so great that it would undermine not only the Indonesian economy but ultimately contribute to Suharto stepping down as president in May 1998.[34]

When the British and French operators fled the country during the May 1998 riots in Jakarta (which had been instigated by the Asian economic crisis), PAM Jaya regained control of the operation and refused initially to relinquish authority when the crisis passed. With Suharto out of power, they contended that the contracts executed at the behest of the former president violated public tendering rules and thus were invalid. Ongoing labor protests in opposition to the concession to the private companies also undermined service and tariff collections over the next two years. By 2000, however, the management challenges from labor had been resolved, and with no political will on the part of the national government to back the local government's desire to undo the arrangement, new contracts with the two companies were executed.[35]

In the years following the acceptance of water supply delivery in Jakarta as a private concession, the two companies requested repeatedly to increase water rates. One increase, a 40% hike, was challenged by the Jakarta Consumers Community (Komparta). The Central Jakarta District Court ruled in favor of the consumer group in 2004 and ordered that the increased rates be suspended until, as the court put it, the water companies "can provide better services and proper information to the customers about their operations."[36] Yet even with some constraints on their company's ability to raise rates, this did not translate into pressure to provide better service for everyone. As Argo and Laquian note, because the concessionaires received compensation based upon the volume of water delivered rather than on revenue collected, there was an incentive to extend water lines to more affluent and more heavily demanded customers and to limit the extension of piped water to poor communities that had more difficult problems of access and thus lower water usage.[37]

The concession agreements had the ambitious (and ultimately unrealistic) targets to achieve universal service by 2022, to reduce water losses to 26%, and that the water would be potable at the tap.[38] In the immediate future, the coverage was supposed to increase to 75% by 2008, but later, it was lowered to 64% even though it never reached 60% coverage. In the two decades following the beginning of the private water concession, access to piped water in Jakarta increased modestly from 44.5% to 59.4% but this reduced target was not reached until early 2019. In the meantime, a citizens group, calling itself the Coalition of Jakarta Residents Opposing Water Privatization (KMMSAJ), challenged the legality of the concession agreements themselves, in part because of failure to reach the targets set, in part because of the rate increases that occurred even though coverage and quality targets remained unfulfilled, but also more fundamentally about the right of a private company to control drinking water. On March 24, 2015, the Central Jakarta District Court ruled that Jakarta's water privatization was illegal, that it violated human rights to water, and that the water system should be returned to public control. Although a higher court granted the appeal of the private operators, in October 2017, the Indonesia Supreme Court confirmed the initial ruling that Jakarta's water privatization was illegal.[39]

The return of control over water service to Jakarta residents through PAMJAYA came with the expectation that it would seek to expand its customer base to serve a greater share of the population and to make good on the intent of the court ruling to affirm access to water as a right. To accomplish this, however, required continued support from external sources, a combination of nongovernmental organizations and donors active in projects to improve access to safe water and improved sanitation, such as through donor assistance provided by USAID's

WASH (Water, Sanitation, and Hygiene) initiative. The need to bring in outside private vendors in the 1990s to expand Jakarta's water service, and the reluctance to invest public funds in improved water conditions might best be explained by the priority given to supporting new development through investments in expanded highways and land reclamation to create new spaces for that development rather than water and sanitary infrastructure. Although water was a topic of concern in the 1990s, the government's focus was on the "waterfront," land development not investing in access to safe drinking.

Throughout its recent history, inadequate attention was given to what happened to the water that the city consumed and where it was deposited as wastewater. Except for the European enclave in colonial Batavis, Jakarta never bothered to install a sewerage system per se, and stormwater always was intended to be managed by allowing it to flow into the rivers and canals. When drainage and flood control studies were conducted during the 1970s and 1980s, there was always attention given to the volume of solid waste, and how this impacted the flood mitigation efforts because so much of it found its way into the waterways, clogging the floodgates and adding to the debris that reduced the flow capacity. Little attention was paid to wastewater, stormwater runoff and sewerage even though so much of the wastewater flowed from household connections into tributaries or directly into the rivers. Riverfront settlements typically directed the water pipes projecting from their structures into the waterways. As obvious as this was, there was no way to accurately measure how much got deposited. Although the KIP and the Integrated Urban Infrastructure Development Program (IUIDP) brought improved infrastructure to many Jakarta communities during the 1970s and 1980s, Putri notes that "sanitation development including wastewater management was the lowest priority among all elements of IUIDP." Typically, only approximately 5% of the IUIDP expenditures were dedicated to wastewater investments whereas drainage consumed as much as 20% of the funding. There was considerable planning for sewerage and wastewater facilities, beginning with a master plan prepared in 1977. According to Putri, who interviewed the engineers associated with the master plan, "only a small part of the masterplan was implemented, to form the major part of the state sewerage system in 2017."[40] One reason was that the master plan, revised in 1991, 2001, and 2005, took a centralized approach by proposing a sewerage network covering an area of 16,000 hectares. The failure to implement most of suggested improvements was not only related to the heavy costs of putting the pipes underground but also because of disputes between national and provincial governments over funding, the complexity of digging up the city to do the work in the midst of ongoing urban growth, and the lack of local political support because there would be no quick and visible impact associated with such a major public expenditure. Compared to building a new toll road that had a defined and reliable revenue-generating potential and that offered a highly visible response to the traffic problems, paying to bury pipes in the ground garnered little political capital. The neglect of critical urban sanitation needs was an invitation, in the words of Asian Development Bank analyst, Florian Steinberg, to a "ecological disaster."[41]

The World Bank supported water and sanitation projects in Jakarta and other Indonesian cities. Although half of the nation's 230 million residents in 2009 resided in cities, only 1% of this population had connections to sewerage, and Jakarta had the least service among the ten large cities with systems. Little had changed when in 2013 the World Bank released an updated study. Only approximately 1,000 households in Jakarta were connected to the city's miniscule sewerage system. The primary infrastructure to handle sanitary waste was the septic tank, many of which were bottom-less concrete boxes (known as "cubluks") that allowed the wastewater to seep into the ground. Because of the crowded housing conditions in the kampungs, the practice was to place the septic tank under the house, making it difficult to desludge and often putting it in relatively close proximity to the shallow wells that the majority of households relied on for their drinking water. There were no national standards for septic construction or local regulations governing septic tank sludge management or disposal. Direct disposal of sanitary waste into the rivers and tributaries, coupled with flaws in the septic systems, undermined groundwater and surface water quality. The inadequacy of wastewater management throughout Jakarta as well as the effects of similarly provisioned communities upstream explains why the rivers that annually flooded were health as well as ecological disasters. The transfer of urban sanitation services to local government under decentralization after 2000 reduced the likelihood that this national problem would get an appropriate national response. During the 1990s, there were over 200 fecal sludge treatment plants built by the national government to deal with the waste from septic tanks. By 2009 and shift to local responsibility, only a handful of these facilities remained operational[42].

Rather than responding to the need to implement additional components of the sanitation master plan revisions, it was the ambitious development scheme, announced by President Suharto in the early 1990s, to fashion a new center for the desired world-class city Jakarta along its historic waterfront that upstaged the call for environmental planning and management decisions to improve access to clean water and sanitary infrastructure. Yet no sooner than the waterfront city scheme was unveiled that Jakarta confronted a series of historic floods that resulted in toxic waters from its rivers and canals carrying sludge and disease into neighborhoods throughout the city.

Notes

1 Heiden, C.N. van der (1990) "Town Planning in the Dutch Indies," *Planning Perspectives,* 5: 79–81; Schagen, F. van, ed. (1967) *Essays in Honour of Professor Jac. P. Thijsse.* The Hague: Mouton, pp. 5–6.

2 Cited in Caljouw, Mark, Nas, Peter J.M., and Pratiwo (2004) "Flooding in Jakarta," paper presented at First International Urban Conference, August 23–25, Surabaya, p. 47; Blommestein, Willem Johan van [1948] *Een federal welvaartsplan voor het westelijk gedeelte van Java N.p.: nn.*

3 Indonesia, Ministry of Public Works and Power (1953) *Pembangunan Kota Baru Kebajoran* [Development of Kebayoran Baru City]. Jakarta: Ministry of Public Works and Power; Silver, Christopher (2011). *Planning the Megacity: Jakarta in the Twentieth*

Century. London: Routledge, p. 85; Mathewson, David Wallace (2018) "Historic Insitutionalism and Urban Morphology in Jakarta: Moving Toward Building Flood Resiliency Into the Formal Planning and Development System," *Journal of Regional and City Planning*, 29 (3): 199.

4 *Pembangunan Kota Baru Kebajoran* (1953), pp. 8, 16, 24–28; Silver, op. cit., pp. 86–87.

5 Kooy, Michelle and Bakker, Karen (2014) "(Post) Colonial Pipes: Urban Water Supply in Colonial and Contemporary Jakarta," pp. 76–81 in Colombijn, Freek and Cote, Joost, eds. *Car, Conduits and Kampongs: The Modernization of the Indonesian City, 1920–1960*. Leiden: Brill.

6 Wirosardjono, S. (1974) "Conditions Leading to Rapid Urbanization in Jakarta and Its Policy Implications," paper presented at a United Nations Conference, Nagoya, Japan, October 28–November 8; Silver, op. cit. p. 92.

7 Watts, Kenneth (1957) *Outline Plan Djakarta-Raya*. Djakarta: Djawatan Perkerdjaan Umum; Bakker, Karen (2007) "Trickle Down? Private Sector Participation and the Pro-Poor Water Supply Debate in Jakarta, Indonesia," *Geoforum*, 5: 857.

8 Watts, Kenneth (1960) "The Planning of Greater Djakarta: A Case Study of Regional Planning," *Ekistics*, 10: 401–405; Watts, Kenneth (1961) "A Planning Study of the Metropolitan Region of Djakarta (Division of Regional and City Planning, Bandung Institute of Technology, February)," in Sendut, H., ed. *1995 Urban Development in Southeast Asia*. Bound typescript in the Institute of Southeast Asian Studies Library, National University of Singapore, pp. 785–791; Franklin, G.H. (1964) "Assignment in Djakarta – a personal view of planning in Indonesia," *Royal Australian Planning Institute Journal*, 2: 229–231.

9 Watts, Kenneth (1960) "The Planning of Greater Djakarta: A Case Study of Regional Planning," *Ekistics*, 10: 401–405.

10 Kurniawan, Kemas, Silver, Christopher, Widyarta, M. Nanda, and Nuraeny, Elita (2021) "Pulo Mas: Jakarta's Failed Housing Experiment for the Masses," *Planning Perspective: International Journal History of Planning and the Environment*, 36, (2): 285–308.

11 Herbowo, Tisnawinata, K., Moochtar, R., and Simonsen, O.C. (1962) Pulo Mas: Project for a Low-Cost Housing District for the Djakarta Municipality Prepared Under the United Nations Technical Assistance Programme. Copenhagen: self-published, October; Silver, op. cit., pp. 106–110.

12 Jakarta, DKI (1966) *Master Plan, 1965–1985*. Jakarta: DKI Jakarta; Silver, op. cit., p. 112.

13 Silver, p. 112.

14 Jellinek, Lea (1991) *The Wheel of Fortune: The History of a Poor Community in Jakarta*. Sydney: Allen and Unwin Australia, p. 12.

15 Ibid., p. 112.

16 Ibid., pp. 130–149.

17 Clauson-Kaas, Jes, Surjadi, Charles, Hojlyng, Neils, Baare, Anton, Dzikus, Andre, Jensen, Henrik, Aaby, Peter, and Stephens, Carolyn (1997) *Crowding and Health in Low-Income Settlements: Kali Anyar, Jakarta*. Aldershot, UK: Avebury, pp. 25–27.

18 Sethuraman, S.V. (1975) "Urbanisation and Employment: A Case Study of Djakarta," *International Labour Review*, 112 (2–3): 199; Sethuraman, S.V. (1976) *Jakarta,*

Urban Development and Employment. Geneva: International Labour Organisation, pp. 38–39.
19 Cowherd, Robert (2002) "Planning or Cultural Construction? The Transformation of Jakarta in the Late Suharto Period," in Nas, Peter J.M., ed. *The Indonesian Town Revisited.* Singapore: Institute of Southeast Asian Studies, pp. 21–22.
20 Indonesia, Ministry of Public Works and Power, Directorate General of Housing, Building Planning and Urban Development (1973) *Jabotabek: A Planning Approach of Its Absorption Capacity for New Settlements Within the Jakarta Metropolitan Region in Cooperation With the Netherlands Directorate for International Technical Assistance.* Jakarta: Directorate General of Housing, Building Planning and Urban Development, April, p. 149; Silver, op. cit., p. 120.
21 Soegijoko, B.T.S. (1996) "Jabotabek and Globalization," in Yeung, Y., ed. *Emerging World Cities in Pacific Asia.* Tokyo: United Nations University Press, pp. 404–406.
22 Soenarno and Sasongko, Djoko (n.d.) "Participatory Planning and Management for Flood Mitigation and Preparedness in the City of Jakarta," Unpublished paper, see 2.2; NEDECO (1973) *Master Plan for Flood Control and Drainage System of Jakarta.* Jakarta; NEDECO (1981) *A Cengkareng Drain System Study*, April.
23 Simanjuntak, Imelda Rinwaty (2010) "Evaluation of the Flood Defense Policy Making Process in Indonesia: The Case of the Eastern Flood Canal, Jakarta, Indonesia." MS thesis, Delft University of Technology, August, p. 43; Soenarno and Sasongko, op. cit.
24 Indonesia, Ministry of Public Works and Japan International Cooperation Agency (1991) *A Master Plan Study, The Study on Urban Drainage and Wastewater Disposal in the City of Jakarta*; Indonesia, Ministry of Public Works and Japan International Cooperation Agency (1996) *A Study on the Comprehensive River Water Management Plan of Jabotabek*; Coyne, Bellier and Sogreah (1979) *A Cisadane-Jakarta-Cibeet Water Resources Plan*; Iwaco, Delft Hydraulics, DHV, TNO, Indah Karya, Waratman & Assoc., Kwarsa.
25 Soenarno and Sasongko, op. cit., see 2.4.
26 Ibid., see 3.1.
27 Lembaga Ilmu Pengetahuan Indonesia (LIPI) (1991) *Water: Proceeding Book of Water, Environment Topic Number One, December 2–5.* Jakarta: LIPI, p. 9.
28 Ibid., pp. 16–17.
29 Argo, Teti and Laquian, Aprodicio A. (2011) "The Privatization of Water Services: Effects on the Urban Poor in Jakarta and Metro Manila," p. 30. www.wilsoncenter.org/sites/default/files/Argo.doc.
30 Porter, Richard C. (1996) *The Economics of Water and Waste: A Case Study of Jakarta, Indonesia.* Aldershot, UK: Avebury, pp. 16–17, 41.
31 Ibid., pp. 91–92.
32 Crane, Randall. (1994) "Water Markets, Water Reform, and the Urban Poor: Results From Jakarta, Indonesia." *World Development,* 22 (1): 71–83.
33 Argo and Laquian, op. cit., p. 232; Japan International Cooperation Agency (2002) *Jakarta Water Supply Development Project.* Jakarta: JICA.
34 Braadbaart, Okke (2007) "Privatizing Water: The Jakarta Concession and the Limits of Contract," in Peter Boomgaard, ed. *A World of Water: Rain, Rivers and Seas in Southeast Asian Histories.* Leiden: KITLV Press, pp. 298–300.
35 Ibid., pp. 301–304, 313, 315–316.

36 *Jakarta Post*, January 30, 2004; Hudionno cited in Argo and Laquian, p. 239.
37 Argo and Laquian, pp. 241–242.
38 Bakker (2007).
39 Atika, Sausan and Aquil, M. Muh. Ibnu (2019) "Jakarta to Take Over Tap Water From Private Firms," *Jakarta Post*, February 12.
40 Putri, Parthiwi Widyatmi (2019) "Sanitizing Jakarta: Decolonizing Planning and Kampung Imaginary," *Planning Perspectives*, 34 (5): 805–826.
41 Steinberg, Florian (2007) "Jakarta: Environmental Problems and Sustainability," *Habitat International*, 31: 359.
42 World Bank (2009) *Urban Sanitation in Indonesia: Planning for Progress*. Field Notes, WSP 2009. Washington: World Bank, pp. 5–6.

4 Return to the waterfront

In 1994, and several years before the demise of the New Order government, President Suharto announced a plan supported by the national government to redirect the growth of the city back toward its historical links to Jakarta Bay. An ambitious, 25-year waterfront city plan, unveiled under the administration of Jakarta Governor Surjadi Soedirdja, involved reclaiming 2,700 hectares of land stretching 32 kilometers across the existing shoreline. Discussions of a waterfront development plan surfaced as early as 1990 under Governor Wiyogo Atmodarminto to provide space needed to support development. The current spatial planning strategy called for slowing development in the peripheral areas south of Jakarta to keep green strategic groundwater recharge sites.[1] The waterfront plan sought to redirect the city's development focus back toward the historic waterfront that for decades had been struggling to retain an economic function and, like comparably situated Asian cities, to take advantage of the city's waterfront location. Artistic renditions of the proposed waterfront city, as shown in Figure 4.1, captured the intention to turn the waterfront into the new central business district adjacent to the refurbished traditional harbor at Sunda Kelapa. Not only would it add new space through land reclamation, but it enabled redevelopment of existing commercial, industrial, and residential areas that had experienced severe environmental stress owing to flooding, deteriorated infrastructure, the incursion of unserviced informal settlements, and an aged building stock that had been allowed to deteriorate for decades for lack of use. There were already plans to build an elevated portion of the inner toll road along the waterfront to improve access. According to the Indonesian Minister of the Environment, Sarwono Kusumaatmadja, "one of the benefits of the reclamation project is that it is in line with efforts to revitalize the city and to maintain environmental management; such as waste industry management, domestic waste, water supply and garbage management."[2]

The waterfront development project involved three contiguous and differentiated development zones spanning 32 kilometers of the shoreline. The western zone extended from Pluit to Kanal Muara and was designated for additions to high-quality residential uses already underway by private developers. The central zone from Pluit to Kota CBD included a new business district, a Heritage Park to capitalize on remaining historic assets (including the already rehabilitated Kota area) and to expand waterfront recreational uses. The eastern zone from

DOI: 10.4324/9781003171324-5

Figure 4.1 Sunda Kelapa rendering.

Source: Sunda Kelapa: a new vision for a historic city. Jakarta Metropolitan City Government, 1995.

the historic core to the Cilincing River provided space for new industrial estates, warehouses, expanded port facilities, and new worker housing to support the operations of the Tanjung Priok harbor.[3] Slowing the seemingly inexorable march of urban development southward toward Bogor into the water catchment zone clearly was also a part of the rationale for rebuilding and expanding the waterfront development area. Promotional materials for the project proclaimed that it would enable Jakarta to join the ranks of other leading Asian cities, such as Singapore, Hong Kong, Shanghai, and Sydney, that had achieved global stature by revitalizing their waterfronts.

The dingy backwater that Jakarta's waterfront had become by the 1990s did not reflect well for a city with global ambitions. Jakarta's waterfront transformation over the previous century had been the antithesis of neighboring Singapore's development story. But there were several local precedents and development success stories that affirmed the potential of coastal development and that explain Suharto's interest in revitalizing the waterfront. One precedent came when Indonesia's first president, Sukarno, proposed the development of Ancol in the northeast corner of Jakarta as a waterfront recreational area as "a showcase for the nation." Ancol's development began in 1967 through reclamation of 552 hectares of land carried out under a contract with a French construction company and managed by the local company Pembangunan Jaya, headed by a young entrepreneur, Ciputra. The result was a popular beachfront amusement area with docking facilities. On the western edge of the coastal area, Sukarno also created a water catchment area at Pluit to manage flood conditions. Nearby the Pluit reservoir in the 1980s, a local development firm (PT Dharmala Intiland) began the construction of a gated waterfront community equipped with docking spaces for all of the residents.[4] The success of Pantai Mutiara inspired Jakarta's now established real estate developer, Ciputra, to add to the Ancol project, in the 1990s, a 1,160-unit private housing known as Pantai Indah Kapuk. As Kusno points out, "by the mid-1990s, the exemplary works of Pantai Indah Kapuk and Pantai Mutiara had convinced the government of the value and technical possibility of developing Jakarta Bay." The Pantai Indah Kapuk development had been approved by the Jakarta government without any environmental assessment, a decision that the sitting Minister of the Environment, Emil Salim, regarded as a mistake. But both Dharmala Intiland and Ciputra were among the key backers of Suharto's waterfront city, and there was no official effort to stop their waterfront projects.[5]

It is also important to recognize that the timing of the proposed waterfront reclamation scheme benefitted from an economic boom that generated major new developments throughout the Jakarta metropolitan area. Bumi Serpong Damai (BSD), initiated in 1985, and Lippo Karawaci (LK), launched in 1993, were massive suburban new towns constructed in the western periphery of Jakarta by development consortia connected to Suharto. The Chief Executive Advisor to BSD was Suharto's cousin, Sudwikatmono, and the project financing came from two key New Order allies, the Sinar Mas Group and the Salim Group.[6] Overall during late 1980s and through the 1990s until the economy collapsed, approximately 60,000 hectares, a land area far greater than the existing city, were converted from agricultural to new residential, commercial, and industrial settlements, dominated by more than 17 new towns and predominantly residential developments.[7]

One unique new community built within the existing boundaries of Jakarta during this boom era and that demonstrated the potential of land reclamation along the waterfront was Kelapa Gading. Soetjipto Nagaria purchased cheaply 30 hectares of swampy land on the northeastern boundary of Jakarta (near but not on the waterfront) and, after draining the plot, began a gradual build-out on 10 hectares through his development company Summarecon Agung in 1978. Rather than copy the prevailing model of the suburban development, Summarecon built houses in small clusters, adding to Kelapa Gading its first retail market area in 1984 (modeled on the type found in Singapore) and an additional comprehensive market area modeled on the highly successful Blok M in the Kebayoran Baru community in South Jakarta. He added new private schools (which attracted buyers that preferred not to use the Jakarta public schools) and later a food center that attracted food establishments that had an established reputation in other parts of Jakarta. It also built "shop houses" (such as those found in Singapore) which provided spaces for community-scale businesses as well as affordable housing options above. It was the first new town in Jakarta to construct a sports center. Between 1987 and 1995, Kelapa Gading added a new supermarket, and the Kelapa Gading Plaza 1 and Kelapa Gading Plaza 2. The malls were located north of the existing major commercial and governmental complexes close to the waterfront. They proved successful in attracting residents and demonstrated that there was an alternative to the inexorable march of development to Jakarta's periphery.[8] One other distinction of Kelapa Gading was that a substantial proportion of its residents were ethnic Chinese who found the location convenient to their existing businesses in North Jakarta and where the community provided a quasi-gated environment, particularly for some of its residential clusters, from the general Jakarta community where there were definitely antagonisms to this ethnic minority. At the same time, it was also vulnerable to flooding because it had been built upon reclaimed swampland, adjacent to the Sunter River and not protected as other Jakarta communities which benefitted from the remnants of the Dutch era canal system. Extending the flood canal to the eastern communities of Jakarta became a salient political issue in the decades following the 1998 financial crisis and connected to the ongoing debates about how to undertake waterfront development.

The most powerful motive supporting Jakarta's waterfront revitalization was the desire by the government and its development allies to replicate the success of similar coastal land reclamation efforts that were advancing economic vitality in Asian coastal cities. In addition to adding new commercial, residential, industrial, and recreational facilities, the pitch by developers such as Ciputra was that Jakarta's waterfront revitalization would provide the impetus to overall environmental upgrading of the growing city.[9] The Jakarta Waterfront Implementation Board (JWIB) identified five major objectives underlying this ambitious waterfront city project:

1 to revitalize the existing waterfront area to be the new nerve center of Jakarta's economic life;
2 to address the existing problems of North Jakarta such as the constant exposure to flooding, poor water quality, poor drainage, and inadequate transportation infrastructure;
3 to redirect the current growth of Jakarta away from the southern periphery;
4 to provide investment opportunities more in keeping with future urban trends; and
5 to establish a new baseline and precedent for development and management approaches that draws more extensively on the private sector.

The goal of the waterfront development project was to "create new residential and commercial districts in the area which will provide a high quality living environment for upper and middle income families." As Kusno (2013) puts it, the Jakarta Bay mega-project was regarded as essential for the city and the nation, to reposition Jakarta from a city of villages to a world-class waterfront city.[10]

Suharto issued a presidential decree (Keppres No. 52) in 1995 that authorized this public and private partnership. It specified the role of the governor of Jakarta to carry out the necessary land reclamation, to develop the specific plan for the waterfront project, and to oversee implementation of the project through the JWIB. A second presidential decree later in 1995 (Keppres 73) specifically called for reclamation of Kapuknaga Tangerang to serve as a tourist area.[11] The developments on the reclaimed land would be situated in a carefully engineered landscape of pristine waterways, state-of-the-art infrastructure (not currently available anywhere in Jakarta), and abundant open public spaces. Under the presidential decree, 4,000 hectares of reclaimed land was set aside for the Kapuknaga Beach Tourism Development Area at the request of the president's ally, Liem Sioe Liong, one of the powerful development interests in Jakarta backing the initiative. This was done in the face of opposition from the localities directly impacted by the project as well as the government's environment ministry.[12] Besides the set aside for the Salim Group, Suharto's daughter, Siti Hutami Endang Admingsih, received authorization to develop 500 hectares of the reclaimed land through her company, PT Manggala Krida Yudha.[13]

In addition to the new developments opportunities proposed for the reclaimed land, the JWIB would guide revitalization of the existing sections of the city lying within the coastal zone, especially the historic areas of Kota Tua (Old City) and the traditional fishermen villages and maritime activities surrounding the port facilities at Sunda Kelapa. The project envisioned a transportation interchange that included a new light rail line linking the new central business district on the north coast and then extending to the middle-income community of Kebayoran Baru in South Jakarta. An elevated roadway connected the existing and future toll road segments, elevated so as not to require the level of displacement that would result from surface construction but also to avoid creating a barrier between the existing city and the expanded waterfront. Additional green spaces would be provided along the shorefront and on the banks of the rivers feeding into the area to "enhance the ambiance and overall desirability of the development."[14] Lacking in the concept of the waterfront city, as Figure 4.2 suggests, is any accommodation for new spaces for the current working class population.

The role of the JWIB was to coordinate and streamline the development processes by handling the design and regulatory processes to ensure developers of "one roof service."[15] The Jakarta government would contract with private investors to prepare the land for development, install modern environmental infrastructure, and oversee revitalization of the heritage areas of "Kota" and "Sunda Kelapa."[16] The backing from the national government, as affirmed by the two presidential decrees, put the project on a fast track, and the JWIB quickly began to initiate the work.

One key feature of the project was to redevelop the still functioning wooden boat harbor area of Sunda Kelapa and the adjacent community areas boasting some remaining structures dating back to the 17th century. Overall, this represented an area of ten hectares that still accommodated 15,000 inhabitants most of whom found employment in the active harbor either with the fishing vessels or that

Figure 4.2 Waterfront city rendering.
Source: Jakarta waterfront city: rebirth of Jayakarta. JWIB, Jakarta, n.d.

carried shipments between Jakarta and outer islands of Kalimantan and Sulawesi. The proposal for the Sunda Kelapa component of the waterfront city involved a mix of preservation of historic sites and water-oriented cultural and recreational facilities to expand tourism in this area. The conservation components included upgrading the Sunda Kelapa quay as a terminal for passengers and recreational boats, with the commercial work of the wooden ships transferred to another site. Rehabilitation of the historic sites surrounding the preserved Dutch statehouse (repurposed as the city's museum) in Fatahillah Square, refurbishing historic buildings, upgrading the area adjacent to the central railway station (Stasiun Kota), and restoring buildings lining the Kali Besar were key components of the plan. Restoring the Kali Besar areas, one of the main business streets in the 18th and 19th centuries, would add a further attraction to this once vibrant area.[17]

The waterfront city project contemplated a completely new water and sewerage system to serve both existing and new developments and to create the desired ambiance by removing pollution from the canals and rivers that passed through this area into Jakarta Bay. The creation of a development firm with public investments, PT Pembangunan Pantai Utara Jakarta (North Jakarta Shore Development Company) would control the reclaimed land and create joint ventures with the city water authority and private firms for supplying water to the reclaimed area.[18] With a new business district based within the waterfront city, the currently decrepit North Jakarta would regain the economic life it had once enjoyed over a century earlier. Renderings of the new waterfront city depicted a cluster of high-rise commercial towers and verdant landscapes, along with elegant residential complexes connected to sparkling waterways, representing a modern version of the water-oriented place that had been Batavia from the 1600s through the early 20th century.

The costs of the project included not only the millions needed for the reclamation but also the human costs of radically reconstructing the areas. The waterfront plan required displacement of thousands of residents from the vast slum areas that had spread throughout the area over the previous century. This included more than 15,000 fishermen living along waterfront. The environmental advocacy organization Wahana Lingkungan Hidup Indonesia (Wahil), along with other conservation groups, voiced strong opposition to the project because of how it would impact the fisherman community as well as the environment that provided their sustenance. Opposition by environmental and community advocates not change or impede the project since the stakeholders who mattered to the government were the developers, and it was their voices that counted.[19]

The process of expanding the coastal area started quickly. Sand began to be dumped in under the guidance of a Dutch engineering firm in 1996, despite concerns among Indonesian scientists and environmentalists that changes to the landscape might affect sea currents and block the river flow, thereby contributing to erosion and flooding. To mitigate the negative impacts to the existing shoreline, and to protect the existing mangroves, the reclaimed land was to begin 200 meters from the shoreline, with a canal formed between the reclaimed area and the old coast to handle elevated waters. Moreover, much of the estimated 200 million cubic meters of fill needed to reclaim land would come from dredged material that had accumulated in Jakarta's 13 rivers. Dredging to get the needed fill also would "deepen the rivers and decrease the likelihood of flooding."[20]

Additional material for land reclamation would come from other public excavation projects, such as the proposed subway to connect the waterfront to the commercial center at Jalan Thamrin and Jalan Sudirman and continuing southward to Kebayoran Baru. Connecting the waterfront project to other key public projects, such as the subway and the river revitalization, was one way to mitigate some of its development costs. The resulting environmental upgrading by using the river material was another side benefit. As Governor Sarwono put it, rivers like the Ciliwung no longer functioned as rivers because of sand accumulation that reduced its flow capacity. Using dredge material as a source of soil for the reclamation project would also serve to deepen river and allow it to flow as it did in the past. In addition, reclaimed land was less expensive than existing city property for developers. Estimates of the land costs for the waterfront area produced by reclamation was roughly one-half the cost of comparable property in the central business district. Although comparable costs savings had been achieved by building in the peripheral areas of Jakarta, continued development on the periphery would simply perpetuate the process of urban sprawl that the waterfront city project sought to counter. As past experience showed, for example, the reclamation done to accommodate the North Jakarta Pantai Mutiara housing development was approximately US$320 per square meter, whereas land prices in Central Jakarta ran at roughly US$3,000 per square meter. Rather than relying on individual projects by Jakarta's leading developers, the waterfront city scheme would enable a more unified approach to the urban expansion process through a planned land development scheme as shown in Figure 4.3. Even with all the safeguards referenced above, there was concern expressed that waterfront development might impact negatively the ongoing efforts to address flooding and the overall quality of the coastal environment. Although some of filling of sand in the harbor was already underway, Governor Surjadi assured the public that development would not begin until an environmental impact assessment was completed, so that the impact could be verified on potential "flood levels, affected performance on the present river system, flood canals and drainage, the flow of contaminated waste, the quality of water at the coastal area, sedimentation

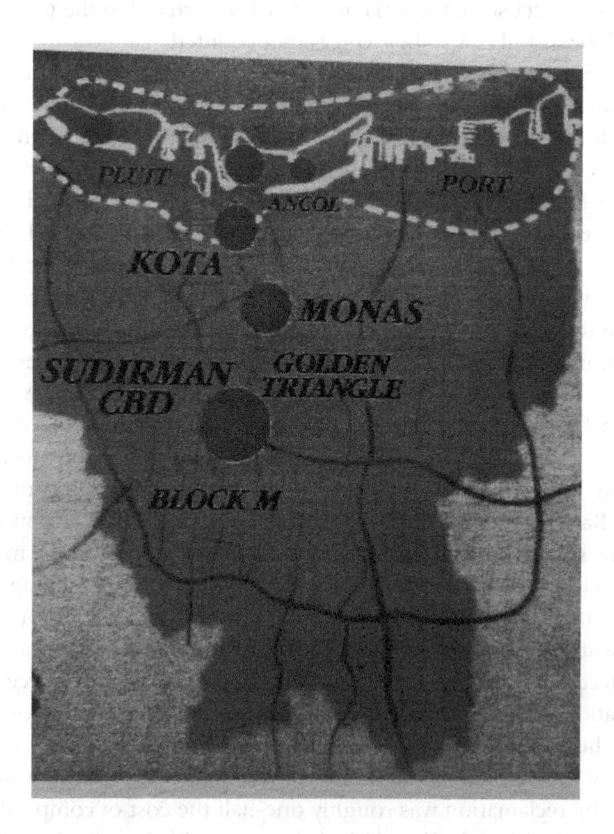

Figure 4.3 Waterfront city scheme and its relationship to Jakarta's major commercial centers.

Source: Jakarta waterfront city: rebirth of Jayakarta. JWIB, Jakarta, n.d.

and erosion of the new coast, tides, and so on." According to Ciputra – one of Jakarta's leading developers and a pioneer in waterfront development in Jakarta – the environmental concerns were not shown to be a problem with comparable development along the coast. What Jakarta proposed in its waterfront city scheme was exactly the sort of development that its neighboring city Singapore had been doing successfully for years, not to mention the Dutch government in the Netherlands and the Japanese in Tokyo Bay. And there was no evidence in those cases of environmental destruction but rather significant urban improvements.[21]

In February 1997, developers announced a massive residential and commercial project for Pademangan, a flood-prone slum neighborhood in North Jakarta. The project necessitated clearing the 190-hectare slum area to be replaced with 18,000 housing units, some for working-class families but most aimed at the middle-class. It included the construction of a million-square-meter shopping center. The new development replaced the semi-permanent and flood-prone buildings with structures designed to withstand flood conditions during the rainy season. A major drawback of the Pademangan project was the lack of alternative housing for those displaced. The national government's target of 110,000 new low-cost housing units for 1997–1998 might meet the need since it was above the previous year's target of 70,000 new units. The problem was that even the more modest goal of 1996–1997 had not been met. So the new development that involved clearance of existing units, as was the case in the Pademangan project, would likely add many more to those in Jakarta lacking a place to live.[22]

By 1997, permits had been issued for nearly 1,000 hectares of reclamation and development of a variety of projects.[23] As the government was finishing up the required environmental assessment of the proposed project by mid-1997, evidence of an emerging financial crisis showed its face as the rupiah began to decline in value. By early 1998, Indonesia's economy was in a free fall. By May 1998, Suharto had resigned, turning over the leadership to his vice president, and marking the end of his over four decades of rule. As the economic crisis deepened, work on the waterfront project stopped but was not officially terminated. In the November 1999 update of Jakarta's Provincial Spatial Plan (RTRW), the reclamation effort continued to be shown in the North Coast Development Zone with the goals of improving and conserving the North Jakarta environmental quality, preserving the fishermen settlement, and enhancing the port and trading function, with the reclaimed land to accommodate international service and trade, housing, and tourism.[24]

Jakarta's Governor Sutiyoso, who had taken office in October 1997 and who would survive the resignation of Suharto in May 1998 and eventually serve under four Indonesian presidents over his ten years in office, sought to keep the waterfront scheme alive. The JWIB remained in place, and Sutiyoso participated in a September 2000 one-day seminar on waterfront cities hosted by the city's Waterfront Implementation Board. In his presentation to the event, he reiterated the mantra that accompanied the initial announcement of the Jakarta Bay megaproject, namely that the project was essential to transform Jakarta into a world-class waterfront city.[25] The waterfront city concept continued as one of a variety of several wishful infrastructure projects for Jakarta's modernization, such as the long-awaited mass transit system. It was now suggested, for example, that when the construction of the subway begins in 2001, it would provide a ready source of earth to create the elevated land upon which the waterfront city project would be built to create a flood proof substrata. While Sutiyoso's words suggested that the waterfront scheme was alive and that the mass transit project was imminent, there was no firm subway construction plans at that time and no commitment to start in 2001. In fact, it would be nearly two decades before it became a reality. The waterfront project no longer had active government involvement although there was some work being done on the ground by some of the developers who had been part of the original scheme. A series of uncoordinated private initiatives happened on some of the land without any active government engagement.

It was not until 2001 when Indonesia appeared to be climbing out of financial crisis of 1998 that the waterfront project received the official green light to resume planning. The city government retained a team of researchers from the Institute of Technology Bandung and Gadjah Mada University to determine whether the project was technically and economically sound, and if it met the "environmental worthiness" test. The environmental assessment was the toughest part of the assignment, and the researcher report went through several iterations until it was ready for review. In August 2001, the Agency for Environmental Impact Control agreed with their assessment that the project was acceptable to proceed, but the official approval never occurred. The agency itself closed down just prior to the massive flood of 2002 that inundated the waterfront communities. The new State Minister for the Environment, Nabiel Makarim, called for a new analysis focused on flood impacts.[26]

Faced with criticism of the North Jakarta project based on what was already underway, especially from environmental organizations but also from local communities that recognized the potential displacements that would result if the full project was put back on track, even the central government seemed to be backing away. In 2003, the Minister of the Environment, responding to critics of the initial decree, declared the proposed reclamation infeasible (Ministerial Decree 14/2003) and recommended that Indonesia's president officially terminate the project. This drew an immediate challenge from a coalition of six development companies who had been part of the project from the outset. The Jakarta government joined forces with the developers and claimed the central government had no power to issue the cancellation. Jakarta petitioned an administrative court to rule against this action. The pro-development forces prevailed in 2004 and then again in 2007, perhaps in response to the impact of the great flood that year (discussed fully in Chapter 5) and new ideas about the relationship between the waterfront initiative and proposed flood control. While the earlier plan was under litigation, the central government was working on a new waterfront plan that included many features of the original scheme but with a greater focus on flood mitigation. The national government challenged the lower court decision and prevailed with the Indonesia Supreme Court in 2009. In yet another twist, the Supreme Court in 2011 reversed its earlier decision and this kept the original project alive.[27]

Throughout the period from 2007 until the 2011 reversal by the Supreme Court, a parallel waterfront reclamation project was under consideration, in part for the same reasons that motivated the 1990s version, that is, to create space for new development within the core area of Jakarta. This objective was complemented by the new and equally compelling consideration, emanating from the post-2007 flood assessment, to provide a grand infrastructure intervention to address the historic levels of flooding and devastation that had become all too common not only along the North Jakarta coast but throughout the city.

While the legal battles over the legitimacy of the waterfront reclamation prevented full-fledged implementation, one key component of the project, cleaning up the rivers that pass through this development zone, continued. Removing the informal settlements that lined the rivers, canals, and the lakes and reservoirs of Jakarta had been identified as a flood mitigation strategy since the early 1990s. Now it was not just about flood control but also about environmental upgrading of a strategic new development zone. President Suharto made what would be his final appointment of a governor of Jakarta in October 1997, Army General Sutiyoso. In one of his final acts as president, Suharto directed designee Sutiyoso to address two of the city's top priorities, to clear the rivers of the informal settlements that had settled there over the past four decades and to reduce traffic congestion on Jakarta's streets. According to the new governor, a first order of business was "to clear the capital's riverbanks of illegal structures."[28] Within a month of taking office, and as the financial crisis was heating up, Sutiyoso announced a plan to remove all of the illegal structures from Jakarta's main river, the Ciliwung, and to transfer the displaced residents to low-cost apartments in East Jakarta. "From the helicopter I saw that some of the riverbanks are still over crowded with slums. I will move the residents off these as soon as possible," he said. To support riverfront removals, he proposed not only the construction of more low-income housing but what he referred to as a "back to nature" program. This involved educating residents concerning the need to adopt an "environmentally friendly approach to development in a bid to minimize the impact of yearly floods" and to encourage an end to illegal dumping in the rivers.[29]

National law prohibited occupation of the riverbanks but a lack of local implementation regulations prevented it from being enforced. So for decades, and in the absence of alternative accommodations, the city allowed informal settlements to occupy the riverbanks and railway lines. This had been Jakarta's de facto low-income housing strategy. Sutiyoso's proposal to clear illegally occupied riverbanks generated widespread protests from those affected and supporting community organizations. A major problem was that the replacement housing (if available) typically was located in outlying areas where land costs were lower, but far from where the riverbank dwellers earned their living largely in the informal sector. As Soenarno and Sasongko noted, when displaced residents "find it too difficult to earn a living in the resettlement areas, they sneak back to their old haunts on the riverbanks."[30] In addition to removing illegal settlers, Sutiyoso indicated support for initiating the East Canal Flood Project to complete the system started in the 1920s. As previously noted, the construction of East Canal to complete the system with the West Flood Canal had been recommended for river water management since the 1970s. Sutiyoso was the first to indicate support for it although this endorsement was not accompanied by a plan for implementation.

Decentralization and the new urban reality

Shortly after taking on the governorship, Sutiyoso confronted an even greater challenge than slum removal and flood mitigation. The Asian financial crisis which began in Indonesia in mid-1997 and peaked by early 1998 sidelined the East Flood Canal implementation and temporarily halted Sutiyoso's plans to push ahead with the informal settlement removals. More important, the fiscal crisis proved to be the precipitating factor in the sudden resignation of five-term President Suharto in May 1998. The onset of the financial crisis in Indonesia was evident in mid-1997 with the free fall of the Indonesian rupiah from approximately 2,500 to US$1 in August to 10,000 to US$1 by December. It would fall further in 1998 (hitting 16,500 to US$1 at its lowest) in part because of the failure of the Indonesian government to make its international loan payment and because of continued flight of capital from Indonesia. The government's financial problem triggered the intervention of the International Monetary Fund (IMF) to offer a bailout but with strings attached. To get IMF assistance, the Indonesian economy needed to remove many of its protections of domestic enterprises, the toughest and most political volatile ones were those that involved eliminating government supports for staples. When the Suharto government eventually accepted the IMF terms and subsequently rice and fuel prices soared (as did other commodities as the value of the rupiah remained depressed), widespread protests broke out throughout the nation, but especially evident in Jakarta. Another virtually uncontested presidential election in March kept Suharto and the Golkar party in office for another term, but generated growing public grumbling when he appointed his daughter and one of his closest friends to cabinet posts and the selection of the B.J. Habibe to the vice presidency. Habibie had served in the Suharto cabinet as Minister for Research and Technology since 1978 when he was selected as vice president in 1998. Growing student protests, which led to the army attack on a group of Trisaki University students protesting outside the national assembly (four of whom died), brought public discontent to a fever pitch.[31] In May, Suharto resigned for stated health reasons and this elevated Habibe to the presidency after just two months as vice president.

Over the previous two decades, the international donor community had been encouraging the Indonesian government to devolve some key central government functions to local and provincial governments under the banner of "decentralization" on the premise that local governments were in a better position to know how to more effectively utilize the funds. The Habibe government saw decentralization as a way to address the economic mess and respond positively to those localities in Indonesia, especially the larger cities, that wanted continued central government financial support (they had little revenue-generating powers) but also more discretion in how to use the funds. At the same time, they also wanted greater taxing powers so that there would be more funds generated locally rather than continuing to rely so heavily on central government decisions. In step with the decentralization movement was the demand for greater governmental accountability through democratic reforms. This included the ability of citizens to elect their leaders and to participate in governance through consultative processes. Within a year of assuming the presidency, the Indonesia government enacted Law 22/1999 and Law 25/1999, the former granting to the local government all responsibilities not reserved for the central government, while latter stipulated that at least 25% of the national budget (not including funds from donors) would be redistributed to the local governments through a transparent distribution process. These new decentralization processes began in 2001.[32] Law 22 and Law 25 were the first two key anchors of the decentralization process that would progress over the next decade and transform Indonesian cities and the surrounding districts into powerful actors in national development. Law 22/1999 granted to the local government responsibilities in health, education, urban services, infrastructure management, agriculture, the environment, and coastal management. Previously, these functions were under the control of national offices located at the local level in a process known as "deconcentration." Under the New Order government's pseudo-decentralization strategy, these "deconcentrated" offices at the local level duplicated the powerless local offices and were the real decision makers. Now these offices and their staff became local government employees. This transfer from central government control to local control of all employees gave the local government the power to allocate staff to meet local needs in a way that had not been possible before.[33] Also, 1999 witnessed the first "free and fair elections" since the 1950s (and before the New Order government). These involved election of members of the local and national legislatures who retained the authority to select the governors, district heads (*bupati*), mayors, and the president.[34] Over the next few years, through a carefully crafted system of extending the opportunity for direct election of local, provincial, and national heads, Indonesia transitioned from a single-party, one-man rule under the New Order to a multi-party democratic system. In 2004, Susilo Bambang Yudhoyono became the first directly elected president. He would go on to serve two terms and help advance the governance reform at the national and local levels.[35] In the case of Jakarta, this meant that when Sutiyoso's two terms as the appointed (and reappointed) governor ended in 2007), the next governor would be chosen through an election. He was succeeded by the first popularly elected governor, his former vice-governor, Fauzi Bowo. Bowo defeated Adang Daradjatun, the former Deputy Chief of the National Police, and became the first governor and one of the rare few government leaders who did not come from a military background.

The direct election of Jakarta's governors, coupled with the expanded responsibilities of the local government and the ability of the electorate to openly assess (and criticize performance without fear of reprisals by the government) influenced how city officials managed critical issues such as provision of clean water, controlling pollution, and mitigating the persistent flooding problems. The tradition of deferring to the central government to respond to urban crises gave way to the expectation that Jakarta's government was the frontline and expected by the electorate to solve the problems. Even in the case of Sutiyoso whose governorship overlapped with the transformation of the government to a decentralized democracy but who was not subject to election during his ten-year term, the mass flood of 2002 (discussed in Chapter 5) put him under extreme pressure to respond to citizen demands for action. Given that Jakarta had not recovered from the worst effects of the Asian financial crisis, the resources needed at the local level to respond not only to the flooding problem but also to implement a new bus rapid transit system exerted pressures that previous governors relied upon the central government to remedy. Jakarta's slum areas continued to expand, in part because of the slow recovery of the Indonesian economy and the continued migration to the city by struggling citizens in search of opportunities. Congested traffic continued to undermine the economic efficiency of Indonesia's main business hub even as Sutiyoso's Transjakarta bus rapid transit system introduced a long overdue modern, affordable, and efficient surface transit system. It was the growing car-owning public that complained about the loss of space on Jakarta's main streets to accommodate the dedicated lanes for the Transjakarta system. Along with continued traffic congestion, the increase in the intensity and regularity of mass flood during the rainy season added another contentious issue for Jakarta's leadership to deal with. This became the issue that could make or break the soon-to-be popularly elected governors of Jakarta. It was not just the flooding that was at issue but its relationship to how Jakarta provided clean water to its residents; how it managed the rivers, canals, and drainage infrastructure, where it allowed urban settlements to locate; and how it managed the necessary reconstruction and revitalization of a built environment erected more than a century earlier. Given the vision of a modern-world city that the New Order regime had advanced through the waterfront city plan in the 1990s and how its development allies kept it in the public eye through their residential and commercial projects in North Jakarta and in the peripheral zones, what was the government going to do to address their part of scheme, namely to clean up the polluted, congested, flood-prone, and environmentally decayed areas that characterized Jakarta. Addressing flooding was key to resolving these larger problems.

Notes

1 *Jakarta Post*, June 29, 1990.
2 Indonesia Property Report (1995), Second Quarter, Jakarta, pp. 30–31.
3 "Expert Upbeat on North Jakarta Reclamation Project," *Jakarta Post*, November 1, 1996.
4 Salim, Wilmar, Bettinger, Keith and Fisher, Micah, "Maladaptation on the Waterfront: Jakarta's Growth Coalition and the Great Garuda," *Environment and Urbanization ASIA*, 10 (1): pp. 63–80.
5 Kusno, Abidin (2011) "Runaway City: Jakarta Bay, the Pioneer and the Last Frontier," *Inter-Asia Cultural Studies*, 12 (4): 519–520; Kusno, Abidin (2013) *After the New Order: Space, Politics, and Jakarta*. Honolulu: University of Hawaii Press.

6 Evawani, Ellisa (2014) "The Entrepreneurial City of Kalapa Gading, Jakarta," *Journal of Urbanism: International Research on Placemaking and Urban Sustainability*, 7 (2): 130–151. doi: 10.1080/17549175.2013.875056.
7 Firman, Tommy (1998) "The Restructuring of Jakarta Metropolitan Area: A 'Global City' in Asia," *Cities*, 15 (4): 229–243.
8 Evawani (2014), op. cit.
9 "City to Set Up Firm in North Jakarta Reclamation Plan," *Jakarta Post*, February 4, 1997; "Firm to Manage Reclamation Project Set Up," *Jakarta Post*, April 19, 1997.
10 Jakarta Waterfront Implementation Board, *Jakarta Waterfront Development Program*, unpublished report, February 27, 1997, p. 2; Kusno, Abidin (2013), p. 102.
11 Salim, Bettinger and Fisher, op. cit., p. 70.
12 Cowherd, Robert (2002) "Planning or Cultural Construction? The Transformation of Jakarta in the Late Suharto Period," in Peter J.M. Nas, ed., *The Indonesian Town Revisited*. Singapore: Institute of Southeast Asian Studies, p. 27.
13 "Legislators Caution City Administration on Coastal Projects," *Jakarta Post*, April 23, 1997.
14 Jakarta Waterfront Implementation Board, op. cit., pp. 11–12.
15 Indonesia Property Report, op. cit., p. 34; Jakarta Waterfront Implementation Board, op. cit.
16 See *Jakarta Post*, February 9, 1997.
17 Jakarta Metropolitan City Government, *Sunda Kalapa: A New Vision for a Historic City*, n.d.; *Jakarta Post*, October 18, 1997.
18 *Jakarta Post*, April 2 and 19, 1997.
19 *Jakarta Post*, April 27, 1995; Kusno (2013), pp. 116–117.
20 Indonesia Property Report, op. cit., pp. 20–21.
21 Ibid., pp. 29, 34.
22 *Jakarta Post*, January 8 and February 14, 1997.
23 *Jakarta Post*, April 23, 1997.
24 Jakarta, DKI (1999) *DKI Jakarta Provincial Spatial Plan (RTRW)*. Jakarta: Regional Development Planning Board; DKI Jakarta Provincial Regulation Number 6/1999.
25 Kusno (2011), pp. 513–531; Kusno (2013), p. 101; Nubianto, Bambang (2003) "Sutiyoso to Pursue Reclamation Plan," *Jakarta Post*, April 12.
26 Hidayat, Agus R. (2003) "Waterfront Tug of War," *Tempo Magazine*, February 4–March 10.
27 Salim, Bettinger, and Fisher, op. cit., p. 70.
28 *Jakarta Post*, October 7 and 14, 1997.
29 *Jakarta Post*, November 3 and 20, 1997.
30 Soenarno and Sasongko, Djoko (n.d.) "Participatory Planning and Management for Flood Mitigation and Preparedness in the City of Jakarta," Unpublished paper, p. 3.2.1.
31 Friend, Theodore (2003) *Indonesian Destinies*. Cambridge, Massachusetts: Harvard University Press, pp. 311–328.
32 Erawan, I. Ketut Putra (2007) "Tracing the Progress of Local Governments Since Decentralization," in McLeod, Ross H. and MacIntyre, Andrew, eds. *Indonesia: Democracy and the Promise of Good Governance*. Singapore: Institute of Southeast Asian Studies, p. 58.
33 Alm, James and Bahl, Roy (1999) *Decentralization in Indonesia: Prospects and Problems*, report prepared for the US Agency for International Development, Jakarta, Indonesia.

34 Sulistiyanto, Bambang and Erb, Maribeth (2009) "Indonesia and the Quest for Democracy," in Erb, Maribeth and Sulistiyanto, Bambang, eds. ***Deepening Democracy in Indonesia? Direct Elections for Local Leaders (Pilkada)***. Singapore: Institute of Southeast Asian Studies, pp. 2–3.
35 Ellis, Andrew (2007) "Indonesia's Constitutional Change Reviewed," in McLeod, Ross H. and MacIntyre, Andrew, eds. ***Indonesia: Democracy and the Promise of Good Governance***. Singapore: Institute of Southeast Asian Studies, pp. 31–33.

5 Job one

Dealing with floods

Flooding has been a routine part of life in Jakarta since its founding as the city of Batavia in the 17th century. Building the city on swampy land adjacent to the estuary created by so many rivers assured it of flooding, especially during the rainy season. City leaders dealt with these inundations through various strategies, such as frequent dredging by manual labor in the early years and later by constructing canals equipped by sluices to divert river waters when they threatened to inundate settled areas. Dams constrained high waters that the canals could not manage; drainage facilities were maintained; and, by the early 19th century, formal settlements moved to higher grounds, thereby abandoning the flood-prone places and lessening the problem for those who could choose where to live. Especially during most of the 300-plus years of Dutch rule, when communities of Europeans occupied the prime locations, these interventions typically satisfied the needs for flood control. Until the flood of 1918 that triggered the construction of a comprehensive flood canal system stretching around some of the outer boundaries of Batavia to divert the waters from the major rivers, flood mitigation was a piecemeal process, a reaction to larger-than-usual floods that did not purport to offer long-term, let alone permanent, protections. Although the 1920s flood canal system remained incomplete until well into the 21st century, initially it seemed that the investment in the West Flood Canal prevented the worst flooding and it was accepted that some level of periodic flooding was inevitable. This infrastructure-dependent approach to flood policy during the colonial regime remained the primary mitigation strategy throughout the late 20th century post-colonial era. Jakarta made little effort to regulate land use in flood-prone areas or to draw upon ecological strategies to manage rainwater.[1] Nor did the city give adequate consideration to the impacts of flooding on the masses of Jakartans excluded from formal communities and forced to reside in the most vulnerable locations along the city's canals, rivers, and their tributaries. Although these residents typically bore the greatest costs from the increasingly intensive floods, the government blamed them for the floods and subsequently displaced them as a flood mitigation strategy. Displacement of the informal waterfront settlement did not stop the flooding, however.

The incidence of major floods, as differentiated from the less severe annual inundations, is of relatively recent origin in Jakarta. The best available evidence

DOI: 10.4324/9781003171324-6

indicates that there were only two major flood events affecting areas adjacent to the Ciliwung in Batavia, one in 1699 and another in 1714. The Ciliwung breached its banks quite severely two more times in the 19th century, one time putting some areas under a meter of water, and then in 1878, following 40 continuous days of rain, all of Batavia and surrounding settlements (*Omelanden*) became a shallow lake. The biggest flood since the founding of Batavia took place in 1918 and instigated the construction of the flood canal project by the government as discussed in Chapter 2.[2] But if one considers the times when the informal indigenous communities experienced significant flooding, severe enough to drive them from their homes for several days and contributing to health problems among those affected, the chronology would need to encompass a list that covers almost every year. Given that where the indigenous population lived was typically in or near to the city's floodplains, it was not surprising that getting wet was a routine occurrence. Not until the flooding levels extended into middle-class communities and shut down commercial zones, which tended to be further away from Jakarta's rivers, did the crisis warrant government intervention.

There were several substantial floods in the 1970s that generated calls to complete the flood canal system that the Dutch had started in the 1920s and to implement additional infrastructure to control the flow of river waters as noted in Chapter 4. Studies were conducted and interventions were recommended to upgrade Jakarta's flood control infrastructure, but few of these were implemented. Memories were short when flooding returned to the annual inconvenience confined to those living along the riverfronts, and so there was no impetus to act. That changed, however, when Jakarta experienced three severe floods between October 1995 and February 1996. Given that these floods resulted in approximately 150,000 houses being flooded and affected approximately one-half million residents, pressures mounted on city leaders to take some serious actions. In the case of the February flood, there were 22 deaths. On February 10, 231 millimeters of rain fell, the highest one-day total recorded in the previous 50 years. It paralyzed the entire city for several days as the waters of five rivers, the Ciliwung, Mampang, Pesanggrahan, Sunter, and Krukut, all passing through densely built-up areas of Jakarta, overflowed their banks. The overflowing rivers sent out a blanket of water between one and two meters deep on many streets. This came on the heels of a slightly less severe flood a month earlier (on January 7) that resulted in 11 deaths. Later research on the flood impact disclosed another fatality, namely the death of the Ciliwung River itself as evidenced by a massive fish kill caused by pesticides filling the river when the floodwaters receded.[3] Pump failures, the exceptional rainfall, and rivers clogged with solid waste all contributed to enabling the 1996 flood to replace the 1918 event as the worst in the city's extended history of floods. In the days following the 1996 flood, as had been the post-disaster response to earlier events, there were renewed efforts to assess why the flooding was so severe and what could be done to prevent a reoccurrence in the future.

Besides the heavier than usual rainfall, a newly cited explanation for the scale of flooding was the loss of green spaces in and around the urbanized area to capture and absorb the heavy rainfall before it flowed into the rivers. Owing to rapid development of Jakarta's suburbs during the 1980s and 1990s, agricultural lands and green areas disappeared. The combined area of land approved for housing development in districts and cities surrounding Jakarta, through which its rivers flowed, was more than 60,000 hectares between 1983 and 1992. Much of this land was in new towns, Bumi Serpong Damai and Tigaraksa in Kabupaten Tangerang (west of Jakarta), Bekasi 2000 in Kabupaten Bekasi (east of Jakarta), and Cariu in Kabupaten Bogor (directly south of Jakarta). Offices, shopping center, industrial estates, and expanded commercial agriculture fueled the development surge. One result was that "large amounts of prime agricultural land ... converted into housing and industrial areas," while the commercial agriculture resulted in an estimated 500,000 tons of sedimentation pushed into the Ciliwung River annually because of erosion. In addition, there were other pollutants generated by the development. Firman and Dharmapatni cited evidence that the rivers flowing through the region, especially the Ciliwung, Sunter, Cipinang, Mookervart, and the Banjir Canal, became highly polluted because of upstream activities. These developments decreased the capacity of the upland region to serve as they have traditionally as a way to capture surface water and recharge the aquifer through water infiltration.[4]

By the mid-1990s, the proportion of Jakarta's total volume of green areas, including the largely uninhabited islands in Jakarta Bay, accounted for just 5% of the 65,000 hectares of land contained within its administrative boundaries.[5] Urban forests that served as water catchments disappeared, especially in the mountain areas south of Jakarta where the construction of resorts replaced the previous agricultural uses. The loss of forested areas added to the volume of runoff into the rivers. In addition, vegetation along the riverbanks that absorbed rainfall disappeared with the expansion of informal settlements. The growing number and size of the informal settlements had the added effect of generated debris to further clog the waterways. As early as 1990, experts warned that all of the essential open spaces would soon vanish if Jakarta and the surrounding governmental bodies failed to take action to preserve them.[6] In the Pluit area of North Jakarta, the flooding was particularly severe since a large swath of its greenery (22 hectares) had been paved over for a new "mega-mall" to serve new luxury development. According to Deputy Governor Tb. M. Rais (speaking on behalf of Governor Soedirdja following the February flood), "the water condition in the city has become worse due to the development process and the only way to cope with the problem is bringing environmental conditions into the equation." He indicated that consideration of water conditions needed to be an important component of the city's revised master plan for 1985–2005, given that the lack of controls over the location of urban development obviously contributed to the recent extensive flooding.[7]

One incremental improvement implemented in advance of the 1995 and 1996 floods was the elevation of some roads in critical locations, most notably the road connecting the city to the international airport. This did nothing to abate the incidence of flooding but merely kept the traffic moving and the airport in operation during less intensive inundations. It remained evident that the protection of the critical water catchment areas of South Jakarta lacked enforcement even though placing development restrictions on the remaining green spaces was one way to reduce flooding. The proposed restriction did not apply to the massive suburban housing, industrial, and commercial developments that were already granted land permits and pertained only to the small-scale resorts and vacation homes built in these restricted areas. A *Jakarta Post* article on February 13, 1996 noted the problem of inadequate open space, suggesting that there "is no longer any way we can go on ignoring the naked reality that compared to other capital cities in Southeast Asia, Jakarta has the smallest area of parks and water catchments."[8] Two days later, the government announced a ban on new projects in the Puncak, the popular resort area in mountainous region south of Jakarta through which many rivers flowed and the area designated in Jakarta's land development plans as the water catchment zone to recharge the aquifer and absorb runoff. This cool highland zone close to crowded Jakarta experienced an influx of resorts, luxury houses, and restaurants over the previous three decades to serve as a getaway spot for city residents. For nearly a decade, the government expressed concerns about the rapid development of Puncak but took no actions to ensure that new structures conformed to the spatial plan intended to safeguard the catchment area. The state minister for national development planning, Ginandjar Kartasasmita, announced that President Suharto approved the 1996 ban "to prevent a repetition of the major flood that wreaked havoc on Jakarta" a week earlier. Ginandjar announced a cessation of building permits in this area and affirmed that government would demolish structures in violation of the current plan. An additional flood prevention project proposed was the construction of a two-kilometer canal linking the Cisadane and Ciliwung rivers to ease the burden of carrying water in the Ciliwung. Another was to increase requirements for new construction projects to maintain additional green space. Early in 1997, the Jakarta government promulgated a new rule requiring all high-rise buildings in the city to provide green space, most likely by greening their roofs rather than maintaining as green the expensive land surrounding these structures.[9] A more ambitious proposed measure involved dredging Jakarta's rivers and to remove the garbage that clogged them and contributed to sedimentation. It was reported that 11,000 cubic meters of trash had to be removed from Jakarta's floodgates following the flood and that this buildup of debris clearly contributed to the rivers overflowing their banks. Governor Surjadi expressed determination "to tighten control of building activities and people occupying riverbanks." But there was no mention of planning better alternatives for Jakarta's poor to reside.[10] As time would reveal, however, none of these proposed interventions progressed beyond the announcement phase.

Cleaning debris from the Ciliwung had long been a goal advocated by Jakarta's small but determined environment group, so the connection between the historic flooding and restoring the Ciliwung resuscitated a dormant program of river restoration. At the urging and the enthusiastic endorsement of Environmental

Minister Salim, the Indonesian government launched the national "Clean River Program" (Program Kali Bersih, or PROKASIH) in 1989. PROKASIH was created in conjunction with the new environmental enforcement agency, Badan Pengendalian Dampak Lingkungan (BAPEDAL), an agency intended to address pollution in rivers throughout the nation. Initially, BAPEDAL targeted 20 rivers for monitoring compliance in 8 of the 27 provinces. This included three of Jakarta's rivers, the Cipinang, Ciliwung, and Mookervaart, as well as three others in the surrounding metropolitan area, namely the Cisadane, Citarum, and Cileungsi. The strategy behind PROKASIH was to set up a river water monitoring system to assess the levels of pollution and water flows and to use these data to strengthen regulation of industrial and nonindustrial discharges into the waterways through the authority of environmental ministry.[11] Another component of PROKASIH was widening the riverbeds to improve water flows. Urban development along the rivers, especially along the Ciliwung where it ran through Central Jakarta, was just five meters wide, a reduction from its original 15 meters width. The resulting reduced flow capacity contributed to its flooding potential during the rainy season. As the longest river passing through the heart of the city, the catchment area along the Ciliwung supported densely populated informal settlements numbering in excess of one million inhabitants. These settlements, lacking access to formal sanitation services, added their household waste to the other contributors to the river's pollution.[12] Besides controlling dumping into the river, particularly by industrial facilities, PROKASIH supported a plan to increase the river width from 20 to 60 meters where the Ciliwung passed through South Jakarta and simultaneously to regreen the riverbanks. PROKASIH endorsed removal of informal settlements to facilitate regreening, while critics alleged that the real motive was to create space for new high-end development, rather than to achieve environmental objectives or to mitigate flooding.[13] The river widening and river cleanup plans advanced through PROKASIH proved more ambitious than the short-term interventions that followed the 1995 and 1996 floods and required extensive financial investments and political will, both of which were lacking. The proposed clearing of the riverfront informal settlements and dredging to widen and deepen the Ciliwung remained as priorities as the challenges of flood management intensified over the next decade.

The virtues of PROKASIH as an initiative to reduce river pollution were a featured topic in a UNESCO-sponsored Jakarta seminar in December 1991 entitled "Water, Environmental Topic Number One." The opening speech by the director of the UNESCO regional office for Science and Technology, Dr. J. Hillig, identified how extensively Jakarta relied on groundwater for its needs because of the river pollution that PROKASIH identified as a problem to solve. He noted there were "tens of thousands of shallow wells in people's back yards." In addition there are 3,000 registered wells that went down at least 100 meters in the ground and drew water from the deeper aquifer. This resulted in a dropping of the groundwater level between two and three meters since the 1970s and contributed to a noticeable increase in land subsidence in the center city. Hillig noted that the "Indonesian authorities are aware or these problems and are working on solutions," most notably building "more water treatment plans in order to reduce the dependency on groundwater." Citizens could contribute to reducing the problem, he pointed out, by following the recommendation of the City Public Works Agency to "build water catchment pits in order to increase the infiltration of rainwater in the rainy season and thus replenish the groundwater body."[14] In the 1995 and 1996 floods that occurred four years later, there was no connection made in the post-flood assessments between the issue of land subsidence, river water pollution, and overreliance on groundwater that contributed to greater incidence of flooding than had been the normal annual problem. It would take future, and more catastrophic, flood events to make it clear that surface water pollution and overreliance on groundwater was a recipe for disaster. In the meantime, there was no expanded water treatment capacity created or used, and the call for citizens to help replenish the aquifer never progressed beyond a good idea.

In fact, concerns about flooding and its mitigation faded from public discourse quickly once the dry season commenced later in 1996. The promulgated Puncak development ban was not enforced and new developments continued there without any regard to impacts on the water catchment zone or potential contributions to runoff into the rivers. There were no demolitions undertaken to restore green areas in the water catchment zone and no green roofs appeared on Jakarta's many new high-rise buildings despite what continued to be a construction boom throughout 1996 and into early 1997. The Cisadane to Ciliwung canal construction remained at the "good idea" stage but without any move by the national government toward implementation.[15] Jakarta's Deputy Governor Tb. M. Rais admitted that the city's planning needed to "match with the city's water condition, an evaluation that should have been much earlier." The new State Minister for Population and the Environment, Sarwono Kusumaatmadja, concurred with the need for better planning but maintained that it was climate change that had contributed to the heavier than usual rainfall not inadequate mitigation efforts, so flooding was inevitable. As he put it, "even if the urban management in Jakarta had been better, the kind of floods we had were going to happen anyway."[16] But that did not preclude the potential benefits from follow-up actions such as improved drainage, moving people from the riverbanks, managing the water pumping stations better, and halting the upstream developments to lessen problems downstream. Indeed, as he seemed to imply, there were interventions that could have been undertaken but were not.

If the prognosis following the 1996 flood was accurate, future floods of a similar magnitude were likely. But what had not been predicted was the onset of a fiscal crisis in Indonesia beginning in mid-year 1997, as evidenced by the plunging value of the Indonesian rupiah, the subsequent fiscal belt tightening by the government, the sudden and unexpected change in the national government administration, and the appointment of a new Jakarta governor. The result was that all measures related to flood mitigation in Jakarta stopped as the nation coped with simultaneous financial and political crises.

Renewed political pressure to deal with flooding in Jakarta occurred in early 2002 and well before Indonesia's political and economic crises had abated fully.

In late January 2002, another flood of historic proportions – the new "number one" historic flood – swept through Jakarta. As observed by the local director of the Church World Services relief organization in a situation report to the home office on January 31,

> Torrential rains resulting in serious flooding have swamped the greater Jakarta area over the last five days, leaving the city paralyzed. The water has reached up to nearly 10 feet in some places and has claimed at least 14 lives.

He acknowledged accurately that the 2002 flood dwarfed all previous incidents in terms of the area affected and number of people displaced. Flooding was reported in 20 of the city's 37 districts and forced several hundred thousand people to evacuate their homes.[17] A BBC report put the high floodwater mark at 13 feet in some areas of the city, with "muddy brown water" covering between 15% and 20% of the city. It reported that several unnamed environmentalists – it still was dangerous to openly criticize government – "blamed the flooding on years of bad city planning, which has led to building-work on green-field sites. That has caused more rain to run into the city's many rivers rather than soak into the ground."[18] The rains continued into early February pushing the numbers of displaced higher and expanding the death toll to 50 persons.

West Jakarta experienced severe flooding, with many of the settlements situated along the West Flood Canal turned into "a large swimming pool filled with dirty, stinking water from the waterway" as the rain pummeled the city for "half a day."[19] West Jakarta communities such as Kelurahan Bambu Utara had 385 housing standing in 15 centimeters of water, South Jakarta along Pakubuono Street in Kebayoran Baru 30 centimeters of floodwater, and the East Jakarta communities along the Cipinang River dealt with as much as 50 centimeters of water. Areas of Central Jakarta located on higher ground had up to one meter of floodwater. And this was just a consequence of day one, January 28, of what would be three days of continuous rain. By evening on January 28, conditions deteriorated and one East Jakarta community Cipinang Indah, situated alongside the Cipinang River, registered flooding two meters in depth. And still the rains continued.[20] By January 30,

> more than three-quarters of Jakarta was under water ... Traffic was at a standstill ... People had to turn off their engines and could not do anything in the middle of the traffic jam; the cars were surrounded by water like ships on the sea.[21]

One reason for the especially severe flooding in the eastern and western sections of Jakarta was because of a decision by the city's government on how to handle operation of the floodgate at Manggarai in South Jakarta. The Manggarai floodgate controlled water into the West Flood Canal. City authorities opened it to allow the water of the main stream of the Ciliwung River to flow into the West Flood Canal rather than continue downstream in the lower Ciliwung with its reduced flow capacity. Jakarta's water management team determined that the

Inner City Ciliwung River, separated from the main river body by the Manggarai floodgate, could not handle the volume of water pouring down from the mountainous region to the south. Soldiers were stationed at the Manggarai gate to prevent anyone from opening the gate which would have sent a wall of water into the center city, swamping the main commercial street (Jalan Thamrin), the government center (including the Presidential Palace), and Merdeka (Freedom) Square and onward into the historic Kota area of North Jakarta. This decision meant that neighborhoods along the West Flood Canal, as well as other along the swollen rivers in East Jakarta experienced the full force of floodwaters.[22]

In this new era of growing democratization and expanded civic organizations providing support to previously underrepresented citizens, the political fallout of the 2002 flood was as historic and unprecedented as the flood itself. Previous floods had yielded postmortems and expressed anguish by affected citizens but rarely did this lead to overt criticisms of government. In 2002, four years since the fall of the Suharto regime and in the midst of an emerging transition in Indonesia to a more democratic governance system, there was a growing public awareness that if government failed to protect citizens from calamities of this magnitude, they had the right to express criticisms openly. Jakarta's governor Sutiyoso, appointed to his position in 1997 (before Suharto stepped down), seemed slow to recognize the magnitude of the flood and did not call for evacuations until long after extensive flooding had occurred in many neighborhoods. This brought a wave of criticism from many quarters, especially from residents of the most affected neighborhood. Critics noted that the Indonesia's Meteorology and Geophysics Board had warned of potential severe flooding in January and February, but that no preparations had been undertaken by Jakarta's government in anticipation of what had occurred. Overall, the 2002 flood was determined to have cost approximately 10 trillion rupiah (roughly US$1 billion) in direct and indirect losses, not to mention the 80 fatalities attributed to the floods. There were calls for removal of Sutiyoso from the governorship for the delayed response to the flooding. To blunt the criticism and to demonstrate sensitivity to potential causes of the flooding, Sutiyoso announced that he would have his villa in the Puncak demolished to provide additional space for rainfall absorption and hence reduced runoff.

Analysts looking at what caused the 2002 flood claimed, as they had in the aftermath of the 1995–1996 floods, that floodwaters accelerated in volume owing to insufficient green space to absorb the rain in the water catchment areas south of Jakarta. It took only a short interlude of no flooding, according to Caljouw, Nas, and Pratiwo, for the "tragedy of 2002" to be almost forgotten.[23] There had been a host of initial postmortems and follow-up studies triggered by the 2002 flood. But none of measures proposed to prevent future flooding on the scale of the 2002 flood progressed beyond the study stage. Interestingly, the government-sponsored land reclamation on the north coast that had been a victim of the 1997–1998 economic crisis was back on track even though it was likely to increase the potential for flooding. At the same time, massive amounts of debris continued to clog all of Jakarta's rivers, and even with the symbolic gesture by Sutiyoso to tear down his country villa (actually only the small shed built on the land came down), the Puncak development trend continued.[24] Little did any of the analysts realize that the 2002 flood was not the historic flood it seemed to be but rather a dress rehearsal for a far greater event five years later.

It is useful, however, to examine the infrastructure responses proposed in the aftermath of the 2002 flood to prevent future catastrophic events since it did prompt the first serious investigation of the flooding problem since the 1970s. With support from the Japan Bank for International Cooperation (JBIC), the Ciliwung–Cisadane River Flood Control Project was unveiled in 2004. During the project study process, another more modest flood swept through Jakarta in February 2004, providing the study team with tangible evidence of several key factors contributing to the recurring problem. The centerpiece of their multi-stage strategy was the construction of a 913-meter diversion channel linking the Ciliwung and the Cisadane, the same canal first mentioned, but not seriously contemplated, in the aftermath of the 1996 flood. The 2004 Flood Control study also proposed channel improvements on the Cisadane covering a 15-kilometer section downstream from the Pasar Baru Barrage to increase its flow capacity. The project team also proposed improvements to the existing West Flood Canal, including 17 kilometers of channel improvements, expanding and upgrading the Manggarai and Karet Barrages, and construction of several new railroad bridges over the canal. Finally, the project designers proposed improvements to the Lower Ciliwung River amounting to 14.5 kilometers of channel improvement. This involved strengthening its banks, building a new barrage on the Ciliwung-Gajah Mada Canal, and rebuilding as many as nine bridges. The total cost of the improvements would run in excess of 1 trillion rupiah (US$100 million). The study also confirmed that the proposed channel improvements necessitated displacing many thousands of residents along the banks of the West Flood Canal as well as along the lower Ciliwung. In defense of the plan, the project designers claimed that the improvements, especially along the West Flood Canal, would likely mitigate flooding that potentially affected approximately 322,500 people in an area of 1,650 hectares. Improvements to the lower Ciliwung would protect an estimated 418,500 people from the effects of flooding.[25]

Concurrent with the flood mitigation study, Jakarta explored both short- and long-term solutions to the problem of inadequate space for solid waste disposal, another contributing factor to the clogging of the rivers and canals and increased flooding. Since 1989, the city had relied upon the 108-hectare Bantar Gebang landfill located east of Jakarta in Bekasi to absorb most of its daily 6,400 tons of solid waste, a site originally designed to handle roughly half that volume. This landfill would continue to be Jakarta's main landfill for the next three decades, growing to an area of 200 acres and a mountain of debris reaching 40 meters in height. A population of 3,000 would make Bantar Gebang home who mined the recyclable materials as a source of income. Since the Bantar Gebang landfill could not handle all of Jakarta's solid waste, efforts were underway in 2004 to identify a new landfill site, the most likely being a site in Kabupaten Tangerang west of Jakarta. A World Bank-funded study verified that much of the solid waste failed to make it to Bantar Gebang, instead being dumped into the rivers. Despite efforts of local authorities to remove regularly the debris, the volume of waste getting into the rivers increased annually.[26]

Sutiyoso was still Jakarta's governor when the next historic flood of 2007 swept through the city. Another mass flooding occurred in late January 2008, and a freakish small tsunami caused when a dam in Jakarta burst in March 2009 (killing 50 people) "together" changed the flood risk paradigm. These floods raised the heat on the Jakarta government to move beyond rhetoric to address the recurring devastation.[27] The dam that burst on the Pesanggrahan River in 2009 had been built by the Dutch in 1933, with an earthen rather than a concrete foundation. After six and one-half decades of service, it gave way as the heavy rains pushed water over its rim. This incident exposed further that a comprehensive effort at water management and infrastructure upgrading was needed urgently. Fauzi Bowo, who had served as deputy governor under Sutiyoso, had been elected to the governorship in August 2007. Although the perennial challenge of mitigating Jakarta's traffic congestion by expanding the Transjakarta busway system remained a policy priority, the political fallout of the 2007 flood and next several inundations elevated flooding to a comparable and competing concern. Two major flood mitigation initiatives discussed for decades were set into motion by Bowo.[28] The most ambitious initiative was the construction of the East Flood Canal to complement the West Flood Canal, and that would complete the system begun in the 1920s by the Dutch to protect the urban core. The other involved implementing the river improvement project through a comprehensive dredging effort. This would require clearing the surface waterways of the informal settlements that according to analysts contributed to the polluted and clogged waterways. This was by far the most controversial and contested flood mitigation intervention because it would require long-term residents to vacate their homes without suitable alternative places to live. It also presumed, incorrectly as community advocates maintained, that the river pollution problem was largely attributable to the actions of these residents. It did not take into account that upstream practices had more serious impacts on water than the actions of Jakarta's riverfront settlements. Residents of the waterfront communities also pointed out that the government provided no alternative to using the rivers for their waste. For decades, the city authorities accepted their occupancy along the rivers and canals in lieu of formal low-cost housing with sanitary services that they could not provide. And clearing informal settlements was not new. It had been practiced for decades whenever an alternative development desired the space. Yet under Indonesia's more democratic and less repressive political regime, the communities were in a better position to challenge this practice. The emerging contestation over the river "normalization" process was significant enough to jeopardize future Jakarta governments.

Constructing the East Flood Canal had been a top flood control recommendation as far back as the 1973 Master Plan for Drainage and Flood Control of Jakarta prepared by Dutch consultants in concert with the Ministry of Public Works. There were five separate new studies dealing with drainage and flood control beginning with another Netherlands Engineering Consultants (NEDECO)-led

"Quick Recommendation Study" after the 2002 flood. The flood prompted the decision to begin securing land for the flood canal.[29] In addition, a 2002 workshop organized by the National Development Planning Board (BAPPENAS) considered how to prevent future massive floods. It was determined that a partnership between central and local governments would strengthen cooperation on water management between Jakarta, Bogor, Depok, Tangerang, and Bekasi. The central government through the Ministry of Public Works committed 2.5 trillion rupiah for the construction of the East Flood Canal, with DKI Jakarta putting up an equal amount to cover land acquisitions. President Megawati announced the central government's funding commitment for the Eastern Flood Canal development plan in July 2003 under the Sutiyoso governorship.[30] But why did it take 30 years from the inclusion of this canal project in the 1973 master plan for flood control to be announced by Megawati in 2003? And why did implementation not happen with the national government covering one half of the costs?

Prior to the government's commitment in 2003, Jakarta lacked funds to acquire the land and to construct the canal. Until the 2002 flood, the key government actor, Ministry of Finance, refused to include the project in the national budget. Under the Suharto New Order government, the governor served at the pleasure of the president (rather than being elected to the position). So without financial and policy support from the central government, there was no reason for city to initiate the work. Although the central government's share of the development costs had appeared in the 2003 budget, the next obstacle for Jakarta was opposition from land owners. Because there was no active pursuit of the land needed for the designated route of the flood canal specified in the 1973 plan, over the next three decades, residential and industrial users filled the lands needed. When the government decree in 2003 kicked off the land procurement process by the Jakarta government, 1,500 families whose land was in the path of the canal refused to sell or to leave their land. As a result, 70% of the land needed faced contestation by the current owners and occupants.[31] It would take another decade, and a more devastating flood to bring the East Flood canal to completion.

In the years following the 2002 flood, and with only the usual annual inconveniences of the annual high waters for those living in low-lying and vulnerable areas, studies of flood control continued without any notable interventions. In the final report of the Ciliwung-Cisadane River Flood Control Project financed by the Japan Bank for International Cooperation (JBIC) in July 2004, it indicated that even the improved methods of managing solid waste in DKI Jakarta since 2001 was not enough to correct the problem of garbage in the rivers. The report "estimated that more than 3000 m3 are thrown into the channels every day."[32] The report went on to note that in the West Flood Canal and the lower Ciliwung River, there is "direct solid waste disposed by people in the channels" and that this accumulates on the bottom causing "lower flow capacity." One significant problem discovered by the consultants was that there were no laws or regulations in Jakarta prohibiting dumping in the river.[33] Finally, increased urban runoff beyond the capacity of the drainage system to handle it, and the reduced water retention capacity of the lakes and green areas of Jakarta also figured into the increased incidence of flooding.

The 2007 flood and the sinking city

Unbeknownst at the time, the historic flood of 2002 was just the warm-up, and a warm-up not sufficiently recognized, for the rains and the flooding that came again in 2007. Beginning on February 2 and continuing through February 4, the unrelenting rains brought flooding that displaced from their homes an estimated 500,000 residents and covered 60% of the urban area as shown in Figure 5.1. This was a unique flood event, not the typical result of the annual rainfall rushing downstream from the highlands during the rainy season. In this instance, "a strong monsoon storm coinciding with a high tide overwhelmed ramshackle coastal defenses, pushing a wall of water from Jakarta Bay into the capital."[34] The flooding brought as much as 13 feet of muddy water into some neighborhoods. Overall, there were 76 deaths ascribed to the flooding, most because of drowning and electrocution. An additional large number of flood victims reported respiratory and skin diseases from contact with the polluted waters. Six days after cessation of the rains, there were areas of the city still reporting floodwaters nearly two meters deep. This prevented some schools, shops, and factories from resuming business. The economic losses from the flood topped US$500 million.[35] In the aftermath of the 2007 flood, local officials cited the usual litany of factors that contributed to the flooding, including "bad drainage, building shopping malls in water catchment areas and cutting down trees in the hills south of the city..." that sent debris into the rivers and reduced the capacity to absorb rainwater before it rushed into the urbanized area. Unimpressed by the explanations, affected residents whose remarks found their way into the press, blamed local officials for the lack of preparedness.[36]

The stories of the suffering caused by the 2007 flood were comparable to those remembered from the 2002 flood. What distinguished the 2007 flood from all previous ones was the decidedly more robust and sustained responses from the Jakarta and national governments that continued long after the flood itself. At the national level and under direct orders from Indonesian President Susilo Bambang Yudhoyono, the Coordinating Ministry of Economic Affairs convened a task force to consider possible interventions. They discussed moving the capital to the higher ground in Jakarta's outskirts, or perhaps to another island altogether, or simply abandoning the North Jakarta area that had been so badly flooded. None of these options seemed economically or politically feasible at the time. What found immediate support was a plan to contract with Dutch hydrological engineers to undertake an in-depth study of the causes of the flood and to use that data to devise an appropriate response. A team of Dutch engineers was already familiar with the North Coast area, having been there for a decade to advise on the 1990s waterfront reclamation project. They turned their attention to flood mitigation.

Two Dutch consultants from Deltares, Jaap Brinkman and Marco Hartman, initiated a study of the causes of the 2007 flood. They immediately dismissed the contention that linked the flood to climate change. As with previous floods, this one resulted from the regular rains, those "clear distinctive successive rainstorms," that were the usual cause. Why then were the flood conditions so

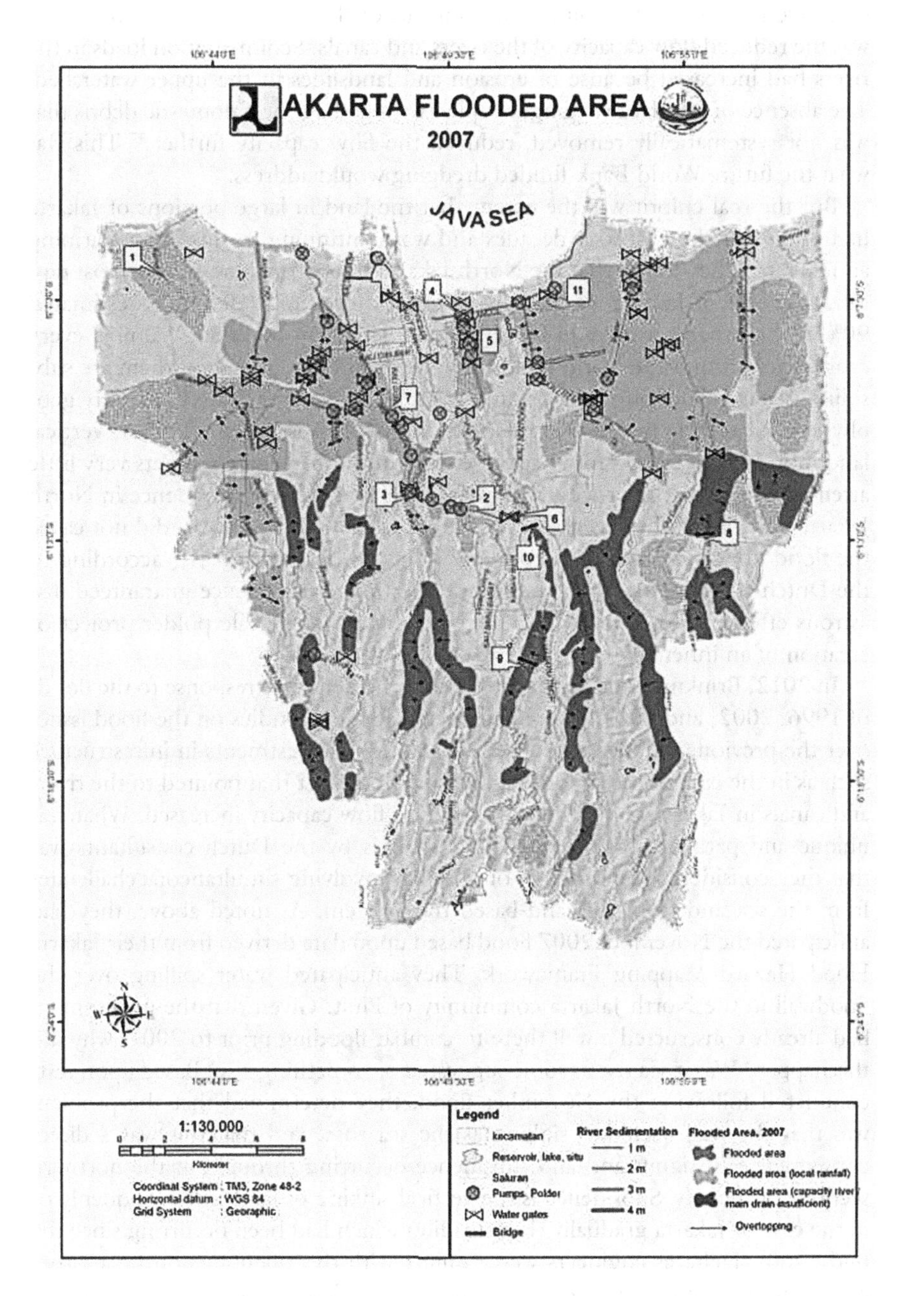

Figure 5.1 Map of 2007 flooded area in Jakarta.
Source: Cipta Karya, Ministry of Public Works, Jakarta.

extensive this time if the rainfall was not out of the ordinary? One explanation was the reduced flow capacity of the rivers and canals. Sedimentation loads in the rivers had increased because of erosion and landslides in the upper watershed. The absence of regular dredging, coupled with garbage and domestic debris that was not systematically removed, reduced the flow capacity further.[37] This was what the future World Bank-funded dredging would address.

But the real culprit was the extent that the land in large portions of Jakarta had sunk over the past four decades and was continuing to sink at an alarming annual rate. They found that the North Jakarta waterfront was sinking most rapidly, that 58% of Jakarta was already below sea level in 2010, and an estimated 90% of the urban area would be in a similar situation by 2030. "Almost every coastal city around the world builds on loose sediment, and all of them are subsiding, regardless of pumping groundwater," says Arizona State University geophysicist Manoochehr Shirzaei, who studies land subsidence. "In fact, vertical land motion is as important as sea level rise, but unfortunately it gets very little attention, because the process is slow." The extent of land subsidence in North Jakarta area, which has been recorded at 12 centimeters per year, did not cause the flood "but increased the depth and duration on some areas," according to the Dutch consultants. More important, continued subsidence guaranteed disastrous effects if not dealt with, either through a large-scale polder project or creation of an inner lake north of the city in Jakarta Bay.[38]

In 2012, Brinkman and Hartman released the definitive response to the floods of 1996, 2002, and 2007. There had been a series of studies on the flood issues over the previous decades, often tied to additional investments in infrastructure such as in the case of the World Bank dredging project that pointed to the rivers and canals in Jakarta as needing to have their flow capacity increased. What was unique and path breaking in the 2012 analysis by the Dutch consultants was that they considered the problem of flooding involving simultaneous challenges from the sea and from the land-based river system. As noted above, they had anticipated the November 2007 flood based upon data derived from their Jakarta Flood Hazard Mapping Framework. They anticipated water spilling over the floodwall in the North Jakarta community of Pluit. Given that the government had already constructed a wall there to combat flooding prior to 2007, why did this happen? Was it sea rise as some suggested or something else? Based upon tests conducted following the November flood, they determined that the problem was that the wall itself was sinking as the sea rose, and that this was a direct consequence of significant land subsidence occurring throughout the northern sections of the city. Subsidence is the vertical sinking of land, either suddenly or in the case of Jakarta gradually (but steadily) which had been occurring since the mid-1960s. Deltares engineers were familiar with this phenomenon since other consulting projects made them aware of the prevalence of subsidence in coastal cities throughout the Southeast Asia region. Using a Piezometer to measure the water level in the groundwater to calculate changes in the surface level, they determined that extractions from the deep groundwater system contributed to Jakarta's subsidence. How much was the city sinking? Comparing data from

1965 when they estimated that subsidence commenced, they found in four test sites that the North Jakarta area had dropped 4.1 meters, another location more in Central Jakarta had dropped 2.1 meters, and two sites further to the south dropped between 1.4 and 0.25 meters. Based upon the current amounts of deep groundwater extraction, and the current rate of subsidence, by 2100, the sinking would reach between five and six meters in North Jakarta and a lesser amount in areas away from the coast. In other words, much of Jakarta soon would be underwater.

There was an alternative. If groundwater extractions stopped within a decade (they estimated by 2020), the overall additional drop could be reduced to just 1.5–2.0 meters. As long as groundwater was extracted, it would continue for sure but at a much slower rate if alternative water sources could be employed. If their assessment was correct, either scenario pointed to the need for prompt and decisive action. Failing to curtail deep groundwater extraction would mean that between four and five million people would need to vacate areas with significant subsidence by 2020. Eliminating, or at least reducing, deep groundwater extraction would buy time but still assume some necessary interventions to protect the city. There was a way, as the report concluded, to turn the current crisis into an opportunity to make long overdue improvements, generate urban redevelopment, and recreate the historic connection to the sea that had been in the public discourse since the 1990s. Their idea was not to retreat from, but rather to advance the city toward, the sea.

The 2012 report proposed the Jakarta Coastal Defense Strategy (JCDS) aimed at simultaneously tackling the river problems, the subsidence issue, and the future threats from climate change and sea level rise. The coastal defense system they proposed, as shown in Figure 5.2, involved creation of three dikes on reclaimed land, one following the current coastline, and two more situated 6 and 14 kilometers, respectively, from the shoreline. Both would be built through land reclamation at sites where subsidence was either minimal or where it had not occurred. Without being subject to continuous subsidence because they were not located upon the aquifer, they could be constructed high enough to accommodate the anticipated sea level rise well into the future. The vast retention basins created between these three seawalls and the mainland would enable potential floodwater to be managed through a pumping system, as well as provide space for a new deepwater harbor to replace Tanjung Priok and a fishing harbor to provide a place for the current fishing communities, likely to be displaced for JCDS construction, to maintain their livelihood.

Of course, the logic of the three dikes strategy followed the example of the Netherlands where a seawall had been protecting a country with much of its land below sea level for hundreds of years. But in the case of Jakarta, constructing a multiple dike system also required other equally ambitious infrastructure interventions on the mainland to make it work. Most important, Jakarta needed a comprehensive water supply system to eliminate the need to rely on deep groundwater for the growing city. As they noted, "land subsidence can only be stopped if deep ground water extraction is replaced by piped water supply." To redirect surface

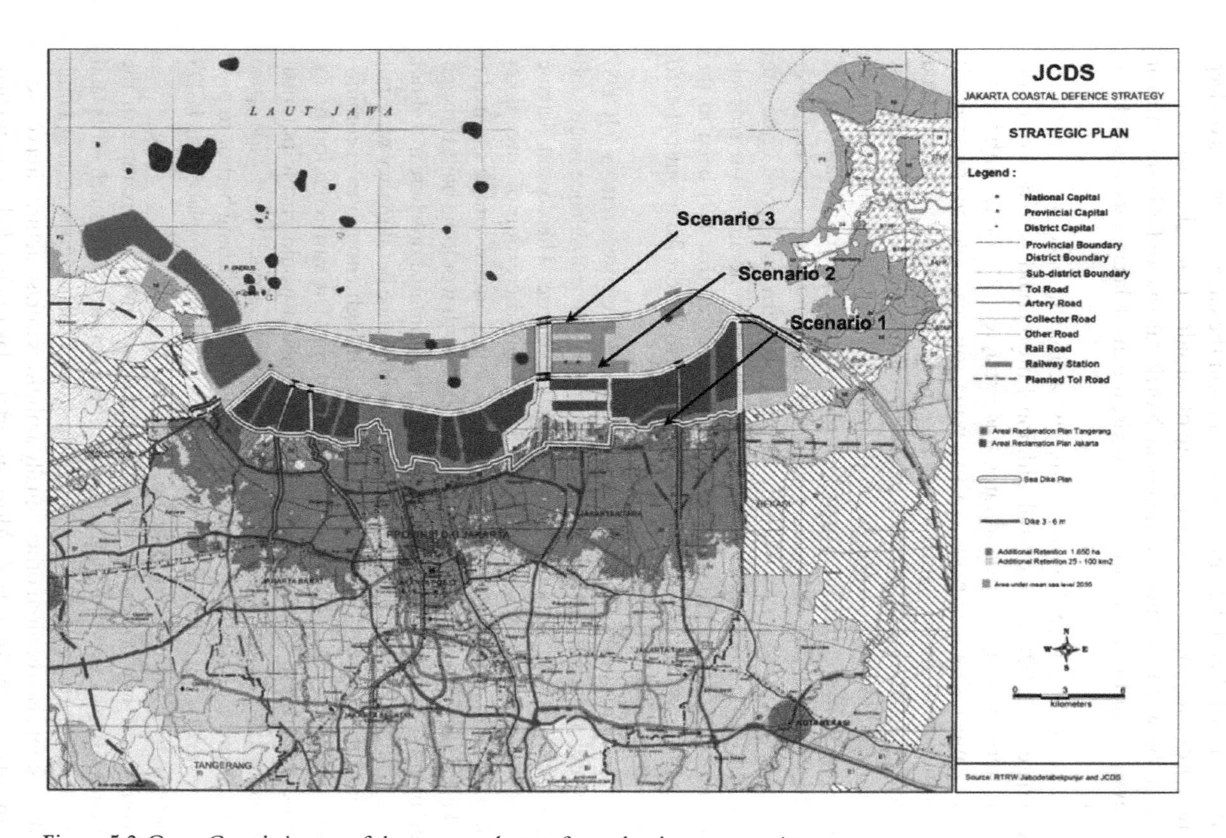

Figure 5.2 Great Garuda image of the proposed waterfront development project.

Source: Mitchell, Beverly (2014) "Can the $40 Billion Great Garuda Project Stop Jakarta From Sinking Into the Sea?" *Inhabitat*, October 16.

waters for domestic, commercial, and industrial uses, the pollution in Jakarta's rivers needed to be removed and improved water purification capabilities added. Cleaning up the rivers also was a precondition to allowing discharge of river water into the retention basins created by the dike system. In their current condition, the rivers would turn the retention basis into a vast open sewer. This necessitated an improved waste management system for the millions of people and thousands of businesses that occupied the river corridors and still used them as their only option for disposing of waste. While this might take time to implement fully, the coastal defense system assumed, largely because of the land subsidence that had already occurred, that initially the rivers would be closed off to the sea. This would become the case over the next few years since additional segments of the floodwall were constructed where the rivers previously flowed into the sea. Already, the city used pumps to carry river waters over the barriers to the sea or by opening sluices.

How to pay for the vast infrastructure improvements that had been neglected by Jakarta over the past half century was probably the most innovative aspect of the coastal defense system. What Brinkman and Hartman envisaged was a system of "livable dikes" and not just barriers to the sea. These dikes would create 3,000 hectares of new developable land for revenue-generating productive urban uses that would help to underwrite much of the costs of development of the dikes and contribute to the cost of cleaning up the rivers, extending the water supply system and improving sanitary and waste management to make possible a reduction of deep groundwater extraction. Without offering details beyond the general concept of the dike system, Brinkman and Hartman sketched the future uses on these "livable dikes" in terms of broad use categories. They suggested that 10% of the land created (roughly 300 hectares) could rehouse the displaced families and shipyard workers who currently resided in the path of the shoreline dike. This amount of land would accommodate up to 60,000 persons but also include a new fishing port, space for ship repair, and facilities for coastal recreation. The inner dike located six kilometers from the coast also included a new connection to the Merak to Surabaya national highway, thereby diverting traffic around rather than through Central Jakarta that was a current source of massive congestion. A new deepwater port located between the inner and outer dikes would replace the outmoded Tanjung Priok facilities in East Jakarta. Railroad lines built on the surface of the dike to support new commercial and residential development would generate land sales for the new city created on the outer dike and offset some of the construction costs.

Their final, and equally compelling, argument in favor of the coastal defense system was that flood control measures without these development opportunities would be too expensive and thereby not done. Coupled with the economic development revenues generated by the livable dikes, the realization of a clean water system, elimination (or at least reduction) of pollution in the rivers and canals, and reduced transportation congestion with new highways and new public transit could transform flood-plagued North Jakarta into an attractive flood-free area that had been the goal for several decades.[39] The project results were fully laid out in a 204-page report distributed by the Ministry of Public Works.

As work on the grand coastal defense system by the Dutch engineers proceeded at the national government level, Jakarta's new governor Fauzi Bowo, elected in August 2007, and in the aftermath of the historic flood, initiated important on-the-ground efforts to address flooding. Bowo was Jakarta's first directly elected governor (all previous governors had been appointed), having served as Deputy Governor under Sutiyoso. He began his governorship understanding the need to act on Jakarta's two most sensitive issues, traffic congestion and flooding mitigation. And he did take action on both fronts. To reduce the volume of cars on Jakarta's congested streets and to provide faster service to commuters, he greatly expanded Sutiyoso's Transjakarta busway system, although this did not allay criticism from those who regarded the dedicated lanes for the transit vehicles interfering with roadway space for their cars and contributing to congestion. It was on flood mitigation that Bowo focused the greatest attention and seemed to generate largely positive press.

The most tangible contribution to flood control involved the construction of the East Flood Canal (shown in Figure 5.3), the unfinished section of the van Breen flood canal system partially built in the 1920s. Completion of the flood canal that had been promised for decades provides protection for the densely populated districts in East Jakarta. The East Flood Canal scheme that the Bowo administration devised differed from the original line in van Breen plan to include a small

Figure 5.3 East Flood Canal completed during the Bowo administration.
Source: Photo by author.

area beyond Jakarta's administrative boundaries in order to touch urbanized areas that regularly experienced severe flooding since the onset of the historic floods in the 1990s. The proposed eastern flood canal swept in an arc 23.6 kilometers in length, passing through 13 kelurahans and terminating at the north coast to the east of Tanjung Priok. Although technically launched by Sutiyoso in 2003 (following the 2002 flood and while Bowo was still Deputy Governor), it stalled for lack of funding and the difficulties of land acquisition. There was widespread community opposition to a route that had been determined without any consultation from those affected by necessary land acquisition. The transition from a centralized to decentralized governance system in Indonesia beginning with the legal changes in 1999 shifted the onus of implementing public works project like the flood canal from the central government to Jakarta. The East Flood Canal was included in Jakarta's 2000–2010 master plan developed under the Sutiyoso administration. Bowo's task was to push the project to completion, to finish acquiring the remaining 407 hectares needed for construction, and to secure additional financial support from the central government to help cover the costs. As Simanjuntak (2010) notes, the transition from appointed to elected provincial leaders like the Jakarta governor under decentralization, coupled with increased press freedom to critically assess the performance of elected officials, likely motivated Bowo (unlike his predecessors) to push the project to completion.[40]

What helped to break the deadlock of landowner refusals to sell was the devastating 2007 flood that likely convinced the owners that the land needed for the 23.57-kilometer canal, and designed to regulate water from six major rivers, the Cipinang, Sunter, Buaran, Jatikramat, Cakung, and Blencong, was worth more to the government than it was on the private market for other purposes. The Bowo administration accelerated the land acquisition for the US$519 million canal. With great fanfare, the Bowo administration announced late in December 2009 that the canal's trench finally reached the sea and just in time for the expected heavy rains of January and February.[41] Governor Bowo pointed out that the canal would reduce the level of flooding by up to 30% although it alone would not safeguard all of Central Jakarta from flooding. Communities such as Pulomas and Cempaka Putih, Bowo noted, still would be subject to flooding unless the drainage system was properly maintained.[42]

Flooding returned on cue in 2008, beginning in the evening of Thursday January 31 and continuing until late in the afternoon of Friday. Although nowhere near as devastating as the 2007 flood, the main roads had knee-deep water and 37 of the 267 subdistricts experienced more than 40 centimeters of water. The toll road to the airport was cut off for a couple of days, and this led to a priority project when dry conditions returned by March to add an elevated road to the airport. Bowo responded to the flood with a set of flood mitigation strategies.[43] Fortunately, the less severe 2008 flood and the milder rains in 2009 (compared to those in 2007) did not generate mass flooding. As a result, Bowo felt justified in awarding a positive self-assessment to the flood mitigation efforts of his administration as he did late in 2009 at the midpoint of his five-year term. He pointed to advancing construction of the East Flood Canal excavation that reached the Jakarta Bay in late 2009, along with dredging in the West Flood Canal and several connecting drainage canals, and introducing into public policy the need for water injection wells in major buildings to replenish the aquifer. Still some critics quickly noted that although the ditch was done, the final construction of the East Flood Canal was far from complete. Moreover, it was noted that the absence of floods in 2008 and 2009 probably had more to do with much lower rainfall and that Jakarta had better brace for 2012 when the next five-year cycle of heavy rains was likely to come.[44]

To Bowo's credit, he revamped the route for the flood canal to extend its reach deeper into the peripheral areas of development than would not have been served by the original path of the van Breen plan of 1918. The settlements of the city had pushed well beyond the original developments of the 1920s and the revised route reached out into Cipinang and Jatinegara in East Jakarta to capture waters from the Sunter, Cipinang, Cakung, Buaran, and Jatikramat rivers to protect an area of 16,500 hectares.[45] This was complemented by design improvements in the canal structure itself. The East Flood Canal was substantially wider than the original West Flood Canal and had slanted 18-meter-wide sides covered by permeable pavement. This allowed water to be absorbed as it rose, rather than pushed quickly along solid vertical solid walls, which was typical of the other canals (Figure 5.4). To prevent new residential settlements and other illegal buildings encroaching on the canal shores, the Jakarta Ciliwung–Cisadane Flood Bureau joined forces with the housing ministry to staff a management board to work with the subdistrict chiefs to regulate construction "to prevent the deterioration of the canal such as Cakung Drain, West Jakarta, where houses were closely located."[46] Another flood mitigation project initiated during the Bowo administration involved the construction of an elevated roadway to ensure access to Jakarta's international airport during periods of high water. Several other initiatives, such as an 800-meter waterway connecting the West and East Flood Canals (when the east canal is complete) and renovation of the Cengkareng Drain in West Jakarta and the Cakung drain in East Jakarta were on the to-do list for flood mitigation.

Improving solid waste management also was critical to reducing the impact of high-water conditions during the rainy season. Establishing a new landfill site and improving waste collection and management was also on the Bowo agenda. According to an official in Jakarta's Public Works department, as many as ten dump trucks worth of garbage is collected daily from the rivers and canals when the flooding is underway, an indicator of how much solid waste was routinely deposited there. Daily clearing of debris from the Pluit Reservoir in North Jakarta, as shown in Figure 5.4, was a necessary flood mitigation strategy. In anticipation of a heavy rainy season, Bowo sought external funding for river dredging to bring relief to communities routinely confronted with flooding.

He noted that many of Jakarta's rivers and canals had not been dredged for decades.[47] The need for comprehensive river dredging had become clear from the vast reach of the 2007 flood caused by overflowing rivers. In April 2008, the

Figure 5.4 Dredging the Pluit Reservoir to remove debris and silt entering from upstream.
Source: Photo by author.

World Bank announced a plan, in cooperation with the Jakarta government, to finance a comprehensive dredging effort in Jakarta's rivers.

The World Bank-financed dredging project was to take place over a three-year period beginning in June 2010 and was expected to reduce the flood-prone areas of Jakarta by upward of 70%. The goal was to return the city to what previously was a cycle of severe flooding once every 25 years, as contrasted with the four major flooding episodes that had occurred over the previous 12 years.[48] The need for dredging had been noted as early as 2002 when the Indonesian government brought in NEDECO to conduct a post-flood reconnaissance study. They found that sediment had reached three meters in the estuaries of the major rivers and drains, and the reduced flow capacity accounted for the overflowing of the rivers during the January and February floods. They recommended that dredging up to ten kilometers from the estuaries of the Ciliwung and Sunter rivers, the West Flood Canal, and the Cengkareng and Cakang drains was absolutely essential or the future would bring even worse flooding. They also reminded Jakarta's leaders that they had recommended as far back as 1973 the need to construct the East Flood Canal.[49] Yet until the post-2007 flood responses under the Bowo administration, there had been no dredging, no new canal, or no efforts to regulate development along the rivers.

The World Bank-funded Jakarta Emergency Dredging Initiative (JEDI) was far more ambitious in scale and would utilize a new method of "mass dredging" rather than the partial approach that was typical of Jakarta's previous efforts. Besides just removing the sediments, JEDI also designated a separate and protected dredge disposal site that would allow for treatment of hazardous soils removed from the rivers. Although most of funding was for the actual dredging, the project included training for local public works personnel and a public education campaign to encourage a change of attitude on how the public treated Jakarta's waterways.[50] Rather than starting in 2010 as initially planned, it launched in 2013 a delay that might have contributed to the victory of Joko Widodo over the incumbent Bowo in the gubernatorial race of 2012.[51]

Jan Jaap Brinkman from the Netherlands-based institute Delft Hydraulics who had headed the analysis of the causes of the historic 2007 flood designed the JEDI project. Brinkman had worked in water and flood management in Indonesia for 20 years and developed a Jakarta Flood Hazard Mapping Framework. This system had accurately predicted the November 2007 flood that, as he contended, was not a result of excessive rainfall but an anticipated high tide that carried seawater back into the sinking North Jakarta coast. In assessing the ongoing flooding problem, Brinkman noted that "the juxtaposition of the high sea tides and the subsidence rate" make approximate four million residents vulnerable to flooding routinely over the next 15 years unless some mitigation efforts were undertaken. Dredging was one key component of flood mitigation along with taking steps to slow down land subsidence.

One obstacle to immediate implementation of the JEDI was the inability of Jakarta to utilize the loans from the World Bank directly, since by law in Indonesia, donor loans must be administered by the central government. The problem was resolved by sharing the responsibility between the Ciliwung–Cisadane Flood Control Office under the Ministry of Public Works and Jakarta's local public works agency, with the central government dispersing part of the funds to the local government to do its part. In the interim, the Jakarta administration did some partial dredging, instituted an early warning system, carried out some maintenance on several strategic dams, and installed mobile water pumps on the toll road to the airport, all of which seemed to mitigate some of the worst effects of a heavy rainfall in February 2009, which actually equaled the volume that had been associated with such severe flooding in 2007. If nothing else, this indicated that even modest improvements in anticipation of flood conditions could make a real difference. Yet to address the long-term effectiveness of Jakarta's water management network of 18 main canals and 500 smaller canals (ranging between 2 and 15 meters in diameter), the comprehensive dredging initiative being undertaken with the World Bank assistance seemed an essential next step.[52]

Delays in financing and final approval prevented the JEDI beginning during the Bowo administration, but the preparation phase identified the settlements along the rivers and canals to be removed to facilitate mass dredging. Many of these communities had existed for decades functioning as the unofficial low-cost housing for the very poor. Lacking wastewater service, they openly used the rivers to meet their domestic needs and in some cases encroached on the waterway. The initiating documents for JEDI prepared by the World Bank in 2009 confirmed that settlement removals would take place to complete the full scope of the project. The initial projection set forth in the World Bank's Integrated Safeguards Data Sheet, which included potential environmental and social impacts, indicated that approximately 5,450 dwelling units (4,200 surrounding a reservoir in North Jakarta, Waduk Pluit) would need to be displaced to accommodate the dredging and rebuilding of the embankments. In the final report on the dredging project released in 2019, it was noted that the resettlements necessary to facilitate the removal of the sediments and waste from the waterways actually amounted to only 328 households, some of whom were relocated in public low-cost housing.[53]

This did not reflect the true social costs of the river improvement project. For example, in one of the sections of the project, the Upper Sunter Drain in East Jakarta, the government demolished 859 structures between November 2014 and April 2015 in anticipation of the dredging that would begin two years later (owing to delays in the project start-up). According to the Land Acquisition and Resettlement Action Plan released in April 2017 (just prior to commencing the dredging), another round of demolitions would be needed, affecting 111 households (and 476 people), but in fact, only 11 units were ascribed to the JEDI project. In fact, many thousands of poor households in Jakarta contributed to the city's flood mitigation efforts as a result of mass clearance of informal settlements, the vast majority of which were not accounted for in the World Bank's figures. These were the social costs of flood mitigation that will be discussed below.[54]

Seeking a transformative intervention

As the administrative complications of the JEDI were getting worked out in 2012, the Indonesia government released the JCDS. Given that the project was unveiled in the year the Jakarta governorship was being contested, it is not surprising that positions on the JCDS were drawn into debates during the campaign. Since what was contained in the report was a concept and not an approved plan per se and because it did not promise an immediate solution to the flooding problem, it was perfectly positioned for Bowo, and his opponent Joko Widodo, the current mayor of the Central Java city of Solo, to fashion their own respective spins on it. Not only did Bowo advocate the plan throughout his campaign but even in defeat, and while still in office, he signed a decree authorizing the beginning of land reclamation called for in the project.[55]

Bowo's challenger, Joko Widodo (popularly known as "Jokowi"), gained national prominence for his success as mayor of Solo in advocating for public engagement and a "populist" style of administering. He brought that style of leadership to Jakarta and went daily into the city's kampungs to build connections with the public. Jokowi's support for the coastal defense plan represented his commitment to an overall environmental improvement strategy for Jakarta. This meant not just enhancing coastal development but also cleaning and restoring the Ciliwung River to its former capacity as the central artery in Jakarta. At times, he referred to the need to clean up all 13 of the city's rivers although no specifics were provided on how to accomplish something that no previous government had even started to do. As he appropriately noted, as did critics of the JCDS, removing pollution from Jakarta's rivers was a necessary precondition for the JCDS to function as intended. Widodo remained an advocate for the JCDS after winning the 2012 Jakarta gubernatorial race and took some steps to prepare the groundwork for implementation. In 2013, the Jakarta government relocated squatters from an area surrounding the Pluit Reservoir in North Jakarta to expand its capacity and transformed part of the cleared area to create a park. It was through incremental steps such as a Pluit Reservoir upgrading that Jokowi confronted the task of actually instituting the river cleanup he promised. It is important to note that Jokowi convinced residents to accept partial removal of the settlement by assuring those displaced that alternative new housing would be available. Rusunawa Waduk Pluit, a complex of 12 four-story apartment buildings with 1,200 housing units, accommodated most of the Pluit displaced residents. As an alternative to the makeshift housing they gave up, the new housing complex offered a solid structure with piped-in water (albeit not always working well). The new park area, as shown in Figure 5.5, created the much-needed green space along the northern coast, a public recreation space

Figure 5.5 Pluit Park created under the Jokowi administration where informal settlement was cleared.

Source: Photo by author.

enjoyed by residents of Pluit and adjacent communities, and it improved the capacity of the reservoir to handle the inflow from the connecting rivers.[56] But the reservoir expansion was not a solution but only a temporary fix. Jokowi also had to contend with a significant flood event in 2013 that was especially severe in affluent areas of North Jakarta situated near Pluit. The heavy flooding in Pluit that also swamped the nearby affluent Pantai Mutiara appeared to justify the necessary improvements to the Pluit and Ria Rio reservoirs.[57] Flooding returned in January 2014 even more severely than in 2013. More than 130 areas of East, West, North, and South Jakarta were affected by between one and four meters of water, forcing over 63,000 people from their homes and disrupting the transportation system for several days. Flooding was a problem throughout the north coastal areas of Java, Sumatra, and areas of North Sulawesi that year. The economic losses experienced as a result of the flooding estimated at 1.87 trillion rupiah. The 2014 flood came as the country was in the midst of a presidential election campaign to choose a replacement for the two-term Yudhoyono. One of the candidates was the newly installed Jakarta Governor, Jokowi.[58]

Since Jokowi won the presidential election in 2014 and given his recent experiences with intensified flooding, central and local government support continued for a scaled-up version of the JCDS that became known as the National Coastal Infrastructure Defense (NCICD). As it seemed likely that the plan was moving toward implementation, opposition intensified, especially from residents of north coast communities supported by the scientific and environmental communities. Critics argued that it was an "outlandish and unnecessary project" that would create a "septic lagoon of trapped freshwater pumped into it from Jakarta's polluted rivers, while at the same time destroying traditional fishing villages and other working-class communities that had enjoyed settlement along the waterfront for decades."[59] During a public consultation of the draft bylaw regulating the spatial planning of the 17 fabricated islets that were the first phase of the coastal defense system (a process of legislative drafting that took more than two years to complete), multiple voices registered opposition to the project. The strongest opposition came from the Indonesian Forum for the Environment (Walhi) that had voiced criticisms about reclamation as far back as the original waterfront plan of 1995. Its deputy director, Zaenal Multaqien, declared that "Jakarta will face an environmental disaster if the city administration continues the land reclamation project."[60] Members of the Indonesian scientific community opposed the project, noting (as the Dutch consultants had pointed out in the 2012 report) that the most effective solution to flooding was to deal with land subsidence in Jakarta by curtailing groundwater extraction. That could only be accomplished if there was an alternative source of clean water coupled with immediate and comprehensive efforts to clean up and stop pollution of surface waters.

By 2015, Jokowi was no longer Jakarta's governor but the newly elected Indonesian president and supportive of the NCICD. In fact, just prior to the induction of Jokowi as Indonesia's president, and the final days of the administration of Susilo Bambang Yudhoyono in early October 2014, the outgoing Economics Minister, Chairul Tanjung, held a hastily assembled groundbreaking

Figure 5.6 Pluit Pumping Station at the sea edge of the Pluit Reservoir in North Jakarta. Source: Photo by author.

for the NCICD at the Pluit pump house in North Jakarta (shown in Figure 5.6 and adjacent to the park Jokowi had built there). Attending were five ministers, one deputy minister and a ranking representative from Jakarta administration. This was the prelude to the official public release of the NCICD plan in December 2014. Jakarta decided that the first action, Phase A, would be to strengthen the existing 94 kilometers of seawalls and river embankments, and to delay the grand strategy of creating the vast outer wall to be used as a new space for urban development. The costs of the floodwall improvements would be shared between the central government, the Jakarta government, and private developers.

A complication with the NCICD implementation was that Bowo had already issued permits back in 2012 to PT Kapuk Naga Indah for the construction of islets C, D, and E, which started to be created by land reclamation offshore nearly as soon as the ink was dry on the original report. This act by the outgoing governor was not preceded, as required by law, by the necessary approvals from the central government. While technically the islets were separate from the process of constructing the giant seawall, the link between their creation and coastal defense project was because all sketches of the potential NCICD showed the islets as part of the grand scheme. By allowing private companies to construct the islets between the shoreline and anticipated outer giant seawall, this had all the appearances of the initial steps implementing the plan. But in 2015, acting-governor Basuki Tjahaja Purnama, known as "Ahok" (who had assumed the office when Jokowi became president), approved development of islet G. The central government took a different view. It denied the permit, indicating it was not within the authority of the local government to issue a permit for this national project.[61] By late 2015, the central government still had not issued the two necessary regulations to enable the NCICD to get started. One regulation involved approval of the master plan itself and the other pertained to the formation of the development authority to oversee implementation. An additional impediment was that the funding needed to begin strengthening the existing floodwalls had not been authorized. The city was supposed to allocate 1 trillion rupiah to start on the eight kilometers of the wall it was responsible for (which was to cost 3.2 trillion rupiah). In fact, the Jakarta government only had 200 million rupiah available in October 2015.

Akhmad Hidayatno and his colleagues, systems engineers at the University of Indonesia who operated a Modeling and Simulation Laboratory, decided to evaluate the potential impacts of the various waterfront redevelopment scenarios that had been advanced over the past several decades to determine if the proposed NCICD represented the best alternative to advance Jakarta's development interests and flood mitigation. They posed the question as a choice between economic betterments or enhancing environmental qualities or perhaps ideally accomplishing some aspects of both. As they determined, the NCICD was the best way to promote economic growth, even if at the expense of environmental conditions, because it called for the reclaimed lands to accommodate an entirely new and modern city in the sea. The NCICD scored poorly on the environmental agenda by contributing relatively little to greening open space objectives and likely to further disrupt maritime commerce. Greening through open space additions without erecting and maintaining the seawall (and without land reclamation) did little beyond add a little green space. Hidayatno contended that greening with select new development accommodated through reclamation in concert with allowances for green open space along the waterfront and protected by a floodwall would be the most fruitful scenario. Without specifying what sort of development would occur on the reclaimed lands, Hidayatno was actually embracing the essence of the original waterfront development plan of the pre-crisis 1990s.[62] It is noteworthy that flood mitigation was not a factor considered in their modeling even though it was a precipitating factor in its initial formulation in 2012. Nor did the model calculate the social costs of the infrastructure approach to flood mitigation.

There was another round of severe flooding in February 2015, hitting not only the usual flood-prone areas in the city but also inundating areas of the peripheral cities of Bekasi and Tangerang. The failure of the pumping system in North Jakarta serving the Pluit polder (blamed on the shutdown of electric power by the utility) was a major factor. Lack of green spaces and failures of drainage again

Figure 5.7 Manggarai floodgate in Central Jakarta clogged with debris from upstream.
Source: Photo by author.

were cited by the minister of public works, and local officials agreed. There was mention of the need for an "eco-drainage" system where water would penetrate into the ground. That was not possible in the rivers lined with a concrete wall that was the embankment design under the JEDI project.[63]

"A river of trash," as the *Jakarta Post* called it, serves as the caption of a photograph of "piles of garbage" blocking the Manggarai Sluice Gate (shown in Figure 5.7) following the rains in April 2018. What the photo shows is that the preponderance of material is dense piles of wood, most likely carried from upstream on the Ciliwung, and not the domestic refuse from Jakarta's riverfront settlements typically cited as the source of the problem.[64]

Notes

1 Octavanti, Thanti and Charles, Katrina (2019) "Evolution of Jakarta's Flood Policy Over the Past 400 Years: The Lock-in of Infrastructural Solutions," *Environment and Planning C: Politics and Space*, 37 (6): 2.
2 Sedlar, Frank (2016) "Inundated Infrastructure: Jakarta's Failing Hydraulic Infrastructure," *Michigan Journal of Sustainability*, 4 (Summer): 1–2.
3 *Jakarta Post*, April 3, 1996.

4 Firman, Tommy and Dharmapatni, Ida Ayu Indira (1994), "The Challenges to Sustainable Development in Jakarta Metropolitan Region," *Habitat International*, 18 (3): 82–86.
5 *Jakarta Post*, May 30, 1992.
6 *Jakarta Post*, July 9, 1990.
7 *Jakarta Post*, February 13, 1996.
8 Ibid.
9 *Jakarta Post*, January 11, 1997.
10 "Jakarta Rivers to Be Dredged to Prevent Floods," *Jakarta Post*, February 15, 1996.
11 Makarim, Nabiel (Deputy I, BAPEDAL) (1991) "Indonesia's Clean Water Program: PROKASIH," in Lembaga Ilmu Pengetahuan Indonesia, *Water: Proceeding Book of Water, Environment Topic Number One, December 2–5*. Jakarta: LIPI, pp. 25–39.
12 Budirahardjo, E. (1991) "Present Rivers Water Quality and the Trend," in Lembaga Ilmu Pengetahuan Indonesia, *Water: Proceeding Book of Water, Environment Topic Number One, December 2–5*. Jakarta: LIPI, pp. 103–108.
13 Nagtegaal, Luc and Nas, Peter J.M. (2000) "Jakarta Greenery: An Essay on Urban Natural Environment," in Nas, Peter J.M., ed. *Jakarta-Batavia: Social-Cultural Essays*. Leiden: KITLV, pp. 263–290; *Jakarta Post*, March 10, 1993.
14 Lembaga Ilmu Pengetahuan Indonesia, Water: Proceeding Book of Water, Environment Topic Number One, December 2–5. Jakarta: LIPI, pp. 9–10.
15 *Jakarta Post*, February 15, 1996.
16 "Ignorance Makes Jakarta Flood Worse," *Jakarta Post*, March 3, 1996.
17 "Jakarta Indonesia Flooding: CWS/ERP Situation Report 31 Jan 2002." Accessed April 13,2020:https://reliefweb.int/report/indonesia/jakarta-indonesia-flooding-cwserp-situation-report-21-jan-2002.
18 "Jakarta Flood Toll Rises," BBC News, January 31, 2002. Accessed January 20, 2020: http://news.bbc.co.uk/2/hi/world/asia-pacific/1791623.stm.
19 Caljouw, Mark, Nas, Peter J.M., and Pratiwo (2005) "Flooding in Jakarta: Towards a Blue City With Improved Water Management," *Bijdragen tot de Taal-, Land-en Volkenkunde*, 161 (4): 455.
20 Ibid.
21 Ibid., p. 456.
22 Ibid., p. 458.
23 Ibid., p. 462.
24 Ibid.
25 Republic of Indonesia (2004) Special Assistance for Project Implementation for Ciliwung-Cisadane River Flood Control Project (I), IP-496. Final Report. Jakarta: Japan Bank for International Cooperation, July; Jakarta Post, May 30, 1992.
26 DKI Jakarta, Dinas Kebersihan Propinsi Daerah Khusus Ibukota Jakarta (2004) *Priority Action Plan for Solid Waste Management in DKI Jakarta, Executive Summary*. November.
27 *Jakarta Post*, March 27, 2009.
28 *Jakarta Post*, January 5, 2010.
29 NEDECO (2002) Final Report Quick Reconnaissance Studies Flood, JABODETABEK, 2002, Main Report, July 10. Jakarta: NEDECO.
30 Simanjuntak, Imelda Rinwaty (2010) "Evaluation of the Flood Defense Policy Making Process in Indonesia: The Case of the Eastern Flood Canal, Jakarta, Indonesia." MS thesis, Delft University of Technology, August, pp. 53–54.
31 Ibid., p. 40.

32 Republic of Indonesia (2004) *Special Assistance for Project Implementation for Ciliwung-Cisadane River Flood Control Project (I), IP-496. Final Report.* Jakarta: Japan Bank for International Cooperation, July.

33 Ibid., p. 36.

34 Tarrant, Bill (2014) "Special Report: In Jakarta That Sinking Feeling Is All Too Real," *Reuters*, December 2.

35 Ibid.

36 "Deadly Floodwaters Recede From Jakarta Homes," February 7, 2007. www.msnbc.msn.com/id/16953369/; Tambunan, Mangapul P. (2007) "Flooding Area in the Jakarta Province on February 2 to 4, 2007," paper presented at 28th Asian Conference on Remote Sensing, November 12–16, 2007, Kuala Lumpur; "Four Metre Floodwaters Displace 340,000 in Jakarta," *The Guardian*, February 5, 2007.

37 Hill, Ed (2013) "Indonesia's Ambitious Plans to Reduce Jakarta Flooding," *Flood List*, December 9. Accessed April 25, 2019: floodlist.com/asia/plans-reduce-jakarta-flooding.

38 Quote from Simon, Matt (2019) "Jakarta Is Sinking. Now Indonesia Has to Find a New Capital." Accessed May 3, 2019: www.wired.com/story/jakarta-is-sinking/?mbid=social_linkedin&utm_brand=wired&utm_social-type=owned.

39 Triple-A Team Consultants, Jakarta Coastal Defense Strategy (September 1, 2010–July 31, 2012) Accessed on April 24, 2019: http://triple-a-team.com/index.php?id=icwrmip-rcmu.

40 Simanjuntak, op. cit., pp. 14–17, 34–35.

41 "East Flood Canal Finally Reaches Jakarta Bay," *Jakarta Globe*, December 29, 2009.

42 *Jakarta Post*, November 22, 2009.

43 Rukmana, Deden (2008) "Jakarta Annual Flooding in February 2008," *Indonesia Urban Studies*. http://indonesiaurbanstudies.blogspot.com/2008/02/jakarta-annual-flooding-in-february.html.

44 Haryanto, Ulma (2009) "East Flood Canal Finally Reaches Jakarta Bay," *Jakarta Globe*, December 29.

45 Febrina, Anissa S. (2007) "Flood Canal Plan Years Out of Date," *Jakarta Post*, March 6.

46 *Jakarta Post*, January 7, 2010.

47 *Jakarta Post*, January 28, 2010.

48 Dredging News Online, April 17, 2008.

49 "Dredging Rivers' Estuaries a Must: Expert," *Jakarta Post*, June 8, 2002.

50 *Jakarta Post*, April 18, 2008.

51 *Jakarta Post*, June 5, 2013.

52 *Jakarta Post*, February 28, 2009.

53 World Bank (2019) *Implementation Completion and Results Report*, IBRD-8210, Jakarta Emergency Dredging Initiative, August 23. http://documents.worldbank.org/curated/pt/153081567169469254/pdf/Indonesia-Jakarta-Urgent-Flood-Mitigation-Project.pdf.

54 Jakarta Water Resources Agency (2017) *Land Acquisition and Resettlement Action Plan (LARAP), Upper Sunter Drain*. Jakarta: Water Resources Agency, April.

55 Hidayat, Agus R. (2012) "Fauzi Bowo paparkan 'Giant Sea Wall'" [Fauzi Bowo explains the Giant Sea Wall], *Kompas*, September 13. Accessed March 15, 2020.

56 Leitner, Helga and Sheppard, Eric (2017) "From Kampungs to Condos? Contested Accumulation Through Displacement in Jakarta," *Environment and Planning A: Economy and Space*, 50 (2): 449; Budiari, Indra (2015) "City Continues Demolition of Homes at Pluit Reservoir," *Jakarta Post*, January 12.

57 Rukmana, Deden (2013) "Jakarta Annual Flooding in January 2013," *Indonesia Urban Studies*. http://indonesiaurbanstudies.blogspot.com/2013/05/jakarta-annual-flooding-in-january-2013-html.
58 Setiawan, Indah and Elyda, Corry (2014) "Losses From Floods Deepen," *Jakarta Post*, January 21; Hussain, Zakir (2014) "Jakarta Rushes to Keep Annual Flooding at Bay," *Straits Times*, January 14; Elyda, Corry and Dewi, Sita W. (2014) "Jakarta Braces for Major Flood," *Jakarta Post*, January 19.
59 Sherwell, Philip (2016) "40bn to Save Jakarta: The Story of the Great Garuda," *The Guardian*, November 22.
60 Wardhani, Dewanti A. (2015) "Jakartans Question Land Reclamation," *Jakarta Post*, October 23.
61 Elyda, Corry (2015) "Court Decision Ends Privatization of Water in Jakarta," *Jakarta Post*, March 24; Elyda, Corry (2015) "Central Government Halts Jakarta Reclamation," *Jakarta Post*, April 13; Elyda, Corry (2015) "A Year After Breaking Ground, NCICD Phase A Yet to Begin," *Jakarta Post*, October 23.
62 Hidayatno, Akhmad, Dinianyadharani, Aninditha Kemala, and Sutriso, Aziz (2017) "Scenario Analysis of the Jakarta Coastal Defense Strategy: Sustainable Indicators Impact Assessment," *International Journal of Innovation and Sustainable Development*, 11 (1): 37–52.
63 Tambun, Lenny Tristia, Rangga, Prakoso, and Marhaenjati, Bayu (2015) "Jakarta Flooded? You Aint Seen Nothing Yet," *Jakarta* Globe, February 10. jakartaglobe.id/news/Jakarta-flooded-you-aint-seen-nothing-yet-officials-say/.
64 "21 Tons of Trash Flows into Jakarta Bay Daily: Research," *Jakarta Post*, December 11, 2018; Boskoro, Yudha (2019) "Bekasi's 'River of Garbage'" *Jakarta Globe*, October 30, 2019.

6 The social costs of flood control

Routinely, post-flood tallies of the costs of the devastation cite the lost revenue from businesses closed or destroyed, the number of Jakarta residents displaced as floodwaters invaded their homes, and deaths owing to tremendous health hazards that accompanied these events. The hardships encountered by those most directly affected in the line of the floodwaters, the riverfront informal settlements, are identified, but not just as a cost but also as a factor contributing to the flooding itself. In the aftermath of the 1996/1997, 2002, and 2007 floods in particular, post-flood commentaries frequently cited the role that these informal riverfront settlements played as a contributing factor to the intensity of the floods. Although themselves victims of the floods, the prevailing view was that they were the problem, that their behaviors, especially using the rivers to deposit their waste and contributing to land erosion by building right on the edges of, and even over the waterways, was the problem. Beginning with the Sutiyoso administration's response to the 1996 flood, and reinforced by the experiences of the 2002 and 2007 floods, removal of these informal settlements was a central component of Jakarta's flood mitigation strategy over the next two decades. This sustained assault on the riverfront communities was accompanied by a public commitment to invest in new low-cost housing for those removed from the path of flooding. However, the promised replacement housing was not always available or necessarily where the displaced persons wanted or needed to reside. Often the community itself challenged the notion that they were the problem that had to be fixed, since the conditions associated with the flooding were not confined to their community but underscored fundamental failures of water management throughout the entire watershed. As the communities knew from past experience, the real reasons for demolishing low-income "slum" communities were often directly related to other motives, typically economic, in order to find available spaces for development in the dense but rapidly expanding megacity.

The removal of kampungs to make way for new development had been a routine occurrence throughout the 20th century as colonial Batavia and then capital city Jakarta pushed the urban footprint into its rural hinterland. Typically, this affected kampungs associated with agricultural districts, such as the kampung clearance associated with the construction of Kebayoran Baru, although many

DOI: 10.4324/9781003171324-7

of the new migrants during the 1950s and 1960s settled along the rivers where there was typically less pressure from new development that preferred higher, drier land. Those that settled along the Cident River in Kampung Kacang were not so lucky. Their community was cleared to accommodate the expanding commercial complex along Jalan Thamrin (see Chapter 3). Informal settlements like Kampung Kacang lacked legal standing and under a 1988 public order regulation (Perda 11/1988) the Jakarta government officially declared "human settlements along railways, along right of ways, along riverbanks, under bridges, and along green paths and parks" illegal.[1] Many were allowed to remain and, in some cases, expand because they occupied land currently not desired for development. Both the government and the residents knew that they could (and would) be removed when there was a demand for their land.

The impetus to aggressively clear the riverfront settlements came with Sutiyoso's appointment to the Jakarta governorship in 1996 (as discussed in Chapter 4) and the command from President Suharto to clear the city's illegally occupied riverbanks. Data from a United Nations Development Program report in 1997 indicated that between 1990 and 1997, there were 68 cases of eviction in Jakarta affecting 194,582 people.[2] Most of these came during Sutiyoso's first year as governor, with 265 separate evictions displacing 108,873 residents.[3] The eviction process ceased during the political and financial crises that began in 1998 and that in general slowed transformation of the community structure of the city that the Suharto government had desired. But the Sutiyoso administration was able to carry out incremental community removals that over his long governorship (1996–2007) actually affected thousands of poor households. Although limited local funds curtailed removals during the first few years following the 1998 financial crisis, this also stymied the construction of the promised replacement housing. Beginning in 2003 and continuing through the end of his term in office in 2007, Sutiyoso made up for the inaction through an aggressive riverfront kampung removal process. According to a report prepared by the United Nations Human Settlements Program, Sutiyoso's governorship between 2003 and 2007 saw over 60,000 families "rendered homeless," a process of removal that often took place, as the report noted, "with violence." In 2003, he authorized the removal of 2,000 families (involving 7,500 people) from the Jembatan Besi community in West Jakarta, 550 people from Sunter Jaya, North Jakarta, and 3,000 people from Kampung Baru. In the latter case, it was not justified as flood mitigation, but to make space for a new housing complex and a shopping mall. Later that same year, hundreds of families were removed from another section of West Jakarta and another 850 homes along the Cipinang River on land designated for the East Flood Canal. Evictions and house demolitions continued through 2007 but at a slightly slower pace.[4]

Along the Cipinang River that meandered from the south through East Jakarta, an informal settlement, Kampung Penas Tanggul, occupied three separate locations on the east and west sides of the river beginning in the 1970s. The reason for this shifting settlement pattern was because this government-owned land first became space for the construction of a waste dump on the west side, and this displaced some of the inhabitants. In 1986, another piece of the community was used for an office building, so again there were evictions and resettlement in a marshy area nearby. The annual floods were a problem the residents dealt with by building their houses above the ground. In 1991, however, the community received a Letter of Eviction from the mayor's office (delivered in paper wrapped around a rock thrown through a resident's window). This time, the *Institute Sosial Jakarta (ISJ)*, a nongovernmental organization that began working with the community in 1986 helped them to organize a protest that successfully stayed for a brief time the order for removal. According to the study by Winayanti and Lang, there were three more evictions (in 1992, 1993, and 1997), leaving just 388 residents by 2000. What the community sought and achieved through assistance from their community organization advocates was an official designation as a place as a Rukun Tetangga (RT). While this did not involve legal title to the lands they occupied, it enables the community to access external assistance, including funds from the US Agency for International Development to construct new community toilets and a septic system and financial assistance from the World Bank to aid local businesses to deal with the ongoing economic crisis in the early 2000s. It also provided residents enough sense of security to undertake home improvements that further solidified their status as a community.[5] They soon had to contend with the floods of 2002 and 2007, however.

Most of the families that endured the worst conditions during the 2007 flood had no alternative but to return to their illegal neighborhoods once the floodwater receded. In 2007, a public order regulation Perda 8/2007 reaffirmed the 1988 regulation that it was illegal to occupy lands on streets within ten meters of a river or other water bodies, railroad lines, and highway flyovers.[6] It was this affirmation of the authority of the local government to remove violators of the setback requirement, unencumbered by anything approximating sound environmental assessment, that sanctioned the mass removals of informal settlements over the next decade.

It was the World Bank's assessment that squatters along the rivers and canals regarded flooding as a routine matter and had no qualms about returning. Without replacement housing within the proximity of the former homes, however, there really was no alternative for them. For that reason, the Bowo administration urged the Jakarta Emergency Dredging Initiative (JEDI) project to begin in squatter-free zones, pending a final decision on how to handle those displaced.[7] As Governor Bowo admitted during the planning stages for the World Bank-funded river improvement process, the displacement necessary to accomplish the dredging and to sustain the efforts to clean up the rivers and canals would be substantial. The government removed approximately 210,000 people from the Ciliwung alone by 2014 to implement the JEDI project. In North Jakarta, there were as many as 150,000 people living in various squatter settlements along the rivers and canals who were potential candidates for displacement as well.[8] Few of these were accounted for in the World Bank project assessment because they were initiated by the Jakarta government. Bowo indicated his intention to work closely with the national government to provide replacement low-rent housing for displaced families currently living in the roughly 70,000 structures that lined the banks of the river. The construction of low-cost housing did not occur in anywhere near the quantity needed because national government funds were not available. Communities such as Kampung Cikini, that were not displaced, continued to rely on the creeks flowing into the Ciliwung to deposit their household waste, as shown in Figure 6.1, since they were not connected to any formal sanitary systems. So the problem of clogged riverways could not be solved just by dredging but required instead a more comprehensive approach.

Although conceived during the Bowo regime, delays in the approval process pushed implementation of JEDI to his successor's administration. When JEDI began in November 2013 under a revised name – Jakarta Urgent Flood Management Project (JUFMP) – it had an expanded scope. It encompassed 11 floodways (involving both rivers and canals) amounting to 67.5 kilometers, and added dredging of four retention basins covering 65 hectares and rehabilitating or reconstructing 42 kilometers of embankments along the floodways and retention basins.[9] It was the rebuilding of the embankments and walls along the rivers and canals that was largely responsible for removal of settlements in 6 of the 15 project sites. In addition to the dredging and embankment reconstruction work, the continuing problem of land subsidence, especially in North Jakarta, resulted in an extension of the project beyond the initial closing date of 2017–2019 and provision of improved pumping capacity in critical areas. As noted in the December 2017 project implementation assessment report, "a new drainage pumping state will be built at the Sentiong-Ancol area, as part of an overall effort to address severe land subsidence that has caused major portions of northern Jakarta to subside substantially below sea levels."[10] The report also indicated that the flooding in the 34 flood-prone *kelurahans* had decreased in depth and duration during the past year, and that the goal of repairing or reconstructing embankments already exceeded the original 42.2-kilometer target by nearly 25%. Still, there was more dredging needed to reach the original target of 25 kilometers.[11]

During the election campaign for the governorship in 2012 and prior to the implementation of JEDI, candidate Joko Widodo had made it clear that his top priority was restoration of the Ciliwung River to its original function as the primary waterway of Jakarta. He pointed to Ciliwung River rehabilitation as the cornerstone of a long-term effort to revitalize all 13 of Jakarta's rivers not just to mitigate flooding but also to enhance the environmental qualities of the city itself. He told voters that his administration would "end evictions, revitalize rather than demolish, *kampungs*, and provide tenure security to neighbourhoods in existence for 20 years or more."[12] It was a promise he and his successor failed to honor, however.

In fact, selective removal of informal settlements took place throughout Widodo's brief tenure as governor (he was elected Indonesia's president in 2014 at the end of the second year of his five-year term, and Basuki Cahanya Purnama ["Ahok"], his vice–governor, took over the interim role). Another massive flood occurred on January 16, 2013, just a couple of months into his governorship, and this contributed to the urgency of pushing ahead with the flood control

Figure 6.1 Creek passing through housing in Kampung Cikini with household drainage.
Source: Photo by author.

initiatives. A dike along the Ciliwung, constructed by the Dutch nearly a century earlier, collapsed and sent a wall of water rushing into the heart of the business district. Deep muddy brown water covered the base of Jakarta's iconic "welcome" statue in the fountain located in the Hotel Indonesia traffic circle. The Jakarta

Globe printed a photo of President Yudhoyono inside the Presidential palace standing in water up to his shins. These images offered a vivid and disturbing reminder that even the previously flood-immune Thamrin-Sudirman business district could be victimized by flooding.[13] In some sections of the city during the 2013 flood, water levels reached three meters. Overall, 97,000 houses were inundated affecting nearly a quarter of a million residents, many of whom had to leave their homes.[14]

Within a week following the January 2013 flood, the Office of the Coordinating Economic Minister announced the Jakarta Coastal Defense Strategy, although this was not going to bring any immediate relief. Even the forthcoming JEDI project would take years to exert any impact. In response to the loss of clean water service in affected neighborhoods, the Indonesia Red Cross (Palang Merah Indonesia or FMI) set up four mobile water-filtration plants capable of processing 120,000 liters of safe drinking water per day. They also sent out 19 water tankers to 500 water points between January 21 and February 17 to distribute over a million liters of drinking water. In addition, they traveled through communities with disinfectants to address the spread of mosquitos, flies, and other harmful insects and rodents to reduce the spread of disease.[15] As with previous post-flood analyses, it was determined that many of the inundated buildings were located on land zoned for water catchments and green areas, but obviously places that had been built upon anyways.

When Ahok assumed the governorship late in 2014, Jokowi's seemingly gentler approach to implementation of the river "normalization" process, as it was now termed, was abandoned. Based on data compiled by the Jakarta Legal Aid Foundation, in 2015 alone, there were 113 separate eviction actions that affected 8,145 families. The next year saw the number of eviction cases increase by another 300, including the entire area in North Jakarta known as *Pasar Ikan* (Fish Market). Ahok was uncompromising in his view that these waterfront settlements were illegal, that they contributed to the problems of flooding as well as being victimized by these events, and that residents along Jakarta's riverbanks needed to be rehoused in a legal situation. When a flash flood along the Krukut River swept through the portions of the affluent Kemang community South Jakarta, Ahok announced that even high-end developments built in violation of the permitted locations (like several that were affected) would need to be removed. In fact, there were no removals of luxury housing in Kemang built in the flood zone of the Krukut. The promise was kept when it came to the riverside communities of the poor. The Ahok administration undertook systematic clearance of *kampungs* along vast stretches of the Ciliwung.

The removal of the informal housing in Bukit Duri, a settlement alongside the Ciliwung in East Jakarta near Jatinegara, typified the swift and uncompromising approach of the Jakarta government under Ahok and euphemistically referred to as "river normalization."[16] Bukit Duri occupied land alongside a winding section of the Ciliwung that routinely flooded. The residents heard that they were in line for removal and tried to preempt the action. They launched a street protest, erected some simple barriers in its narrow streets to prevent heavy construction equipment coming in, and enlisted the support of the media and two local community organizations, the Jakarta Urban Poor Network (JRMK) and Ciliwung Merdeka, to help organize the resistance. As Wilson (2016) noted, resistance by various Jakarta communities to the government settlement-removal process took many forms, including "class suits, street protests and vigils, and physical resistance to removal and neighbourhood beautification campaigns that challenge the characterization of communities as slums."[17] In the case of Bukit Duri, several of these tactics were employed but none proved effective. With assistance from Ciliwung Merdeka, for example, residents documented and mapped the homes in the community that had been constructed and improved to demonstrate the investments that had been made. The Jakarta Legal Aid group assisted them in appealing the decision to tear down the community.[18] The day before the announced demolition in September 2015, residents received their eviction notices. The next day, supported by 600 police and military officers, the demolition crew took down the entire community in just one hour. The government had promised replacement housing for those evicted from the community and based upon the experience of their displaced neighbors in Kampung Pulo who were moved into new housing nearby, they were hopeful. The displaced residents of Bukit Duri found out that their replacement units were located 30 kilometers away from their just demolished East Jakarta community. While perhaps of better quality than the units removed, some of the affected families had lived in Bukit Duri for 15 or more years, but more importantly, this was where their schools and employment were situated.

Also along the meandering Ciliwung adjacent to Bukit Duri, another settlement, Kampung Pulo (shown in Figure 6.2), had met a similar fate one month

Figure 6.2 Pollution in Ciliwung as it passes through Tongkol riverfront kampung.
Source: World Bank Archives, Washington, DC.

earlier. With the help from the same community organizations, they had put up a strong show of resistance to the removal process. But the result was the same – a swift and complete removal of all the housing early in the morning of August 20. The conditions in Kampung Pulo were considered worse during the floods than in Bukit Duri since the community was bordered by the Ciliwung on three sides. But Kampung Pulo had a different history. It had been built and settled during the Dutch colonial era, but as it grew, population pressures pushed residents onto state lands directly on the riverfront. That made the whole community illegal in the view of government. And it was in the path of river dredging project that intended to rebuild the barriers along the Ciliwung. As the JEDI project geared up in 2013, it was widely believed among the residents that the kampung was destined to be removed. In fact, as Hellman and van Voorst noted, the newly elected Governor Joko Widodo visited Kampung Pulo to discuss the potential to be affected by the improvements to the Ciliwung but led them to believe that if they were required to move, they would be compensated appropriately and rehoused. On August 20, 2015, and now under the authority of Governor Ahok, the bulldozers, protected by the police and soldiers from the protesting residents, cleared the community in a day, just as would soon happen in Bukit Duri. Although the occupants on state-owned lands did not receive compensation for their destroyed homes, they were offered a replacement unit in a nearby public low-cost housing tower.[19]

Because of the chance to rehouse nearby, the displaced residents of Kampung Pulo, who accepted the move to the *rusunawa*, had a different experience from those in Bukit Duri. The 2,164 residents who were forced out of Kampung Pulo and accepted the rental flat stayed within the proximity of their jobs and schools. Moving into Rusunawa Jatinegara Barat meant exchanging their unserviced dwellings for secure rental units with substantial amenities although with a higher rent. According to a study by a team of researchers from the Institute of Technology Bandung, the transition by the Kampung Pulo residents into their new rental apartments required a period of adjustment but one that ultimately proved to be largely successful. On the positive side, the community was now free from the annual distress of being flooded during the rainy season nearly every year, sometimes for several months. The Rusunawa tower provided residents with clean water from the public system, a wastewater treatment facility to handle the discharges before entering the communal septic system, and twice daily garbage removal. So no longer was the community's waste being deposited in the Ciliwung or finding its way into the ground to contaminate the community's water supply. In addition, Rusunawa tower provided an on-site health-care clinic, early childhood education facilities, a library, and direct connections to the Transjakarta busway system.[20]

But all of these amenities came with a cost, and the greatest post-relocation challenges were socioeconomic in nature. Now they were required to pay a monthly rent of 300,000 rupiah (US$25) which was significantly greater than that required in the *kampung* housing, and additionally, there were new fees for the water and electric services. Since the majority of the residents earned income from food sales in the *kampung*, their income dropped after relocation because the kampung market had been disbursed. Moreover, the community network that was so important for maintaining social stability in the *kampung* was more difficult to recreate when the residents were stacked onto different floors of the housing tower rather than sharing the ground-level spaces that connected the residents in the traditional community structure. According to a city official asked about the future of low-cost housing in Jakarta, his answer was the "future must be vertical."[21] In effect, some of the problems relieved by the improved physical facilities were countered by other life challenges. But according to the Institute of Technology Bandung researchers, the clear beneficiaries were the children of Kampung Pulo who now enjoyed a more open and environmentally upgraded place to live and play.

Another case of proposed kampung removal from the flood zone led to a different outcome. According to the city administration, the North Jakarta communities of Tongkol, Lodan, and Japat clearly violated the "city administrative regulation on the proper distance between a home and river bank." Kampung Tongkol and its 250 families resided on both shores of *Anak Kali Ciliwung* (a short branch of the Ciliwung which translates as "Child of Ciliwung") but without the required 15-meter setback from the river to facilitate an "inspection street" by the local government. Because of evidence that residents had been using the river to dispose of household waste, coupled with houses that encroached on the waterway, the city informed the community that Tongkol would be cleared. The local community leaders, with support from the Urban Poor Coalition, undertook preemptive actions in advance of the demolition order. The community pointed out to city officials that the 15-meter setback was a radical change from the 5-meter setback that had been required in the 1990s. They requested a return to the previous standard. When the Deputy Governor visited in 2015 to hear their concerns, the community greeted him with refreshments and made a case for the reduced setback while a group of the community youth blocked his departure until there was a tentative agreement to allow for the narrower "inspection street."[22] The community, with the help from volunteering architects and students, designed an acceptable "inspection street" themselves (as shown in Figure 6.3). To do this, residents removed the front rooms of houses that intruded on the riverfront spaces. To compensate for this, some added an upper floor to their small structures. In addition, they removed all of the debris from the river (as well as convincing residents to use bins rather than the river for waste), installed new septic tanks connected to dozens of houses, removed several of the violating structures, and landscaped their newly fashioned "inspection street." Although this action did not fully comply with government specifications, the wide publicity surrounding the Tongkol intervention (assisted by the Urban Poor Coalition coupled with regular visits by academics and journalists to see how the residents had been able to stave off demolition) persuaded the Jakarta government to relent on the previously announced removal.[23] A similar community upgrading strategy was done in Lodan and Japat (located further to the east in the Ancol area) and with similar

Figure 6.3 Tongkol inspection street created by removing front part of bordering houses.
Source: Photo by author.

positive outcomes. By reducing their houses to a single room for sleeping and living and giving up the kitchen, many of the residents now had to cook on the street. The inconveniences of the new living arrangements were compensated for by the pride these community residents felt because they had achieved what other riverfront settlements typically had been powerless to realize, stopping the government's destruction of their homes.[24]

Another nearby community to the north and situated on Jakarta Bay was not so fortunate and got the Bukit Duri/Kampung Pulo treatment. The Pasar Ikan (the historic "fish market") once a thriving center that drew its customers from throughout the city was leveled on April 11, 2016. A force of 4,000 city officials, police and soldiers who backed up a crew with heavy equipment, pushed the community down in less than one hour. Its removal was related to expanding the seawall construction in Muara Baru to protect adjacent areas from storm surge and a continuation of the waterfront revitalization strategy that began with the 1995 Waterfront Plan. Pasar Ikan was long known as a place to get fresh fish in North Jakarta and in the original waterfront scheme called for its revitalization rather than removal. In the meantime, a maritime museum had been installed in one of the nearby historical warehouses, and a modern fish market had been built nearby in Sunda Kelapa to the east. As a result, Pasar Ikan no longer boasted the historic aura it once possessed. It was just a dilapidated informal settlement that impeded revitalization of the north coast.

Low-cost housing was offered to the families displaced by the demolition, and 321 of them signed up for it. The drawback for some was that the housing complex, Rusunawa Marunda, was located 26 kilometers from Pasar Ikan and from the fish market-based employment that the residents still relied on. Although the displaced families were able to live in the new facilities rent free for three months, there was no effort to provide assistance in securing employment near the apartment complex. As one displaced resident noted, he was able to find another job in the fishing business near Pasar Ikan, but it consumed virtually all of his pay for the nearly two hours of transportation to and from the job. Without a job requiring less travel, and when the 151,000 rupiah (plus utilities) begins to be charged, his family (wife, two children, and mother) might need to move from the clean new facility to something affordable.[25] Some of those whose homes were torn down simply stayed, pitching tents and constructing makeshift dwellings while city workers began the process of shoring up the floodwall. As late as early 2018, the site of Pasar Ikan remained a field of littered debris (as shown in Figure 6.4) with a smattering of rickety structures accommodating those that remained. It was also on the ruins of Pasar Ikan where residents of adjacent Luar Batang community were able to hop on the makeshift ferry service to get over to their ***kampung*** whose historic mosque probably explained in part why it was left out of the coastal demolition. Although the area now boasted the maritime museum in refurbished Dutch East India Company (VOC) warehouses, the historic watch tower near the site of the original colonial fort and the new fish market were intended to attract tourism, it was clear that the primary objective of the Jakarta government was to remove another large informal settlement.[26]

Ahok's stern policy of relocating the riverbank dwellers was met with fierce criticism from residents and activities who called his actions "inhuman and against human rights."[27] No doubt the informal settlement removals contributed to his failed campaign for a full term in the 2017 gubernatorial election that he lost to Anies Baswedan, a former cabinet member in the Widodo administration. Two advocacy groups working with kampung residents, the Urban Poor Linkage and the Urban Poor Consortium (UPC), solicited the support of architects, artists, scholars, students, and journalist to work with low-income communities under threat to make them "eviction-proof" by demonstrating strategies for self-improvement. The Tongkol community upgrading was one of their demonstrations. The advocacy groups also used political power to protect the kampungs. According to journalist Evi Mariani Sofian, one of the local advocates,

Figure 6.4 Traditional fish market (Pasar Ikan) torn down along the North Jakarta coast to facilitate the construction of seawall and in anticipation of new waterfront development.

Source: Photo by author.

they assured Baswedan that they would deliver him votes in exchange for recognizing their land right.[28] Baswedan promised a similar community-sensitive approach to flood management that Jokowi had found so successful and as a counterpoint to the heavy hand of the Basuki administration. When reports from Bogor in February 2018 indicated that heavy rains were promising a heavy flow down the Ciliwung to Jakarta, Baswedan went to the Manggarai Sluice "to see whether the water from Bogor was flowing smoothly" and to assure the public that he "was ready to tackle potential flooding as all three sluice gates in the capital had already been opened" and the Jakarta Environmental Agency was clearing out the accumulated 200 tons of debris.[29] Although the JEDI project would still run for another year, the aggressive riverfront removals slowed considerably under the new administration, evidencing that Baswedan upheld his side of the deal. It was also owing to the ongoing efforts of the young professionals working with a growing cadre of kampung leaders to strengthen their skills in participation and creative ways to address community problems.

What did not change was Jakarta's flood mitigation policies based upon the sustained and largely unsubstantiated claims that the kampungs along the

Ciliwung and the other Jakarta waterways were a major contributing factor to city's flood problem. In fact, the kampung residents were victims of a city that decided that providing clean waste, sanitary and solid waste facilities, preserving green spaces to handle stormwaters, and support for infrastructure improvements in the kampungs, many occupied as low-income communities for nearly a century, deserved the same attention as the formal communities. "Living on the margins in poor areas," Thynell observes, "means being exposed to multiple and constituted inequalities that result from centuries of living without public services, or perhaps they result from a deliberate strategy by the government maintain uneven development."[30] One of the key initiatives that the Baswedan government announced in April 2018, and included in the first budget presented to the Jakarta council, was a program to construct 14,000 apartment units, that could be accessed without a down payment (that the city government would cover), for those with a monthly salary between 4 million rupiah and 7 million rupiah. Given that displaced residents from Kampung Pulo and Pasar Ikan faced difficulties in handling rents between 115,000 rupiah and 300,000 rupiah month rents, the Baswedan administration rehousing strategy had no relation to the realities of the very poor. To ensure sufficient budget for this, funding to extend the tap water supply beyond the 50% of Jakarta residents currently having access was excluded from the budget. While it was noted that the transition from private water suppliers to the city-owned Purusahaan Air Minum Jaya (PAM Jaya) had not been achieved, the promise to improve water service to poor North Jakarta communities who never had access to piped water was not regarded as a priority.

Reflecting on what happened in the case of Kampung Pulo is instructive about how once acceptable and marginally supported indigenous communities could be transformed into obstacles to urban modernization. Wilson (2015) notes that Kampung Pulo was always regarded as a "legitimate" neighborhood, as evidenced by being recognized in the local government's administrative structure, where residents paid taxes, and which boasted "inter-generational continuity of residence extending prior to establishment of the Indonesian Republic." Seemingly overnight, it was transformed into a "den of squatters" illegally situated, and one of the contributors to Jakarta's persisting flood problem.[31] Another Ciliwung fronting community, Kampung Cikini, built to accommodate Indonesians who served the affluent households in the adjacent Menteng community, is another case in point. While it has yet to have the bulldozers show up, the spatial plan already shows the future of this densely settled community as eventually transformed to office spaces and other high-end commercial uses because of its central location. Except for the intervention of a group of academic activists to help construct a single set of toilets (one for men and one for women) and a "flyover toilet" (this is a facility built over a stream running through the community which empties directly into the water), a dense section of Kampung Cikini has no other alternative, and there is no movement on the part of the Jakarta government to bring them the same service their neighboring community, Menteng, has enjoyed for nearly a century. Putri (2019) interviewed the local government engineers to find out the status of a 2012 plan to construct a sewerage network by dividing the unserviced areas of

Jakarta into 15 zones so that this monumental task could be accomplished through a more incremental approach. Five years after release of the plan, there was still no political support for it owing to the technical complexities and costs of doing it in such a densely developed city and especially because of how much of the unserviced areas "have been developed informally and incrementally, generating irregular patterns that are seldom congruent with the grid pattern for piping the city."[32] So for Kampung Cikini and other Jakarta communities, as evidenced in the need to utilize a "helicopter toilet" to enable a semi-private form of open defecation (see Figure 6.5), the human and environmental consequences of this service deficiency were significant. And expansion of the water supply to poor communities, intended when the system was turned over to private vendors in the 1990s, brought minimal expansion.[33] Under municipal control, as Bakker contends, the willingness of government to allow water vendors to monopolize service to poor communities provides a revenue source more reliable (and less costly) than extending service directly to poor communities and having them pay, even at elevated tariffs. This is because even with extended service, poor communities would likely rely on shallow wells rather than paying the cost of service since the service itself neither offered better nor more reliable water.[34] The water infrastructure deficit explains the unwillingness by the government to prioritize water service to all citizens as a right without regard to the recipients' ability to absorb the costs of connections.

Figure 6.5 Helicopter toilet in kampung lacking septic or sewerage system.
Source: Photo by author.

It is important to note that clearing these informal and incrementally developed communities and replacing them with orderly and serviced new development has been routinely untaken throughout Jakarta since the 1950s. But what has been built in those cleared spaces are largely housing and commercial facilities to serve the middle and upper classes in Jakarta with much less attention to reconstruct orderly and serviced facilities for the former residents.[35] An alternative approach, drawing upon the tradition of the Kampung Improvement Program from the New Order era involving reconfiguring the spatial structure of kampungs in these vulnerable areas, never has been considered. "While the quality of housing ranges from substandard to acceptable, conditions immediately adjacent to the riverbank are often poor and there is a case for negotiating 'in situ' replacements" according to Dovey et al.[36] Undertaking a rebuilding process sensitive to the design features of the kampung structure but with the opportunity to create buffers and include infrastructure while allowing the existing settlement to persist with a new infusion of housing would mirror redevelopment projects that have proven to be successful alternatives to clearance in other cities.

Kampung activists such as Ciliwung Merdeka (Free Ciliwung) and the UPC, who stood with the residents of Bukit Duri and Kampung Pulo, helped prepare alterative plans that would have achieved the environmental objectives claimed to be intent of the government's intervention, but with less impact on the existing community. As Koh notes, these community advocates "worked with researchers and architects to conduct their own mappings of cultural and environmental landmarks, as well as assessments of safety easement widths along the river that would have the least impact on the kampungs."[37] They offered designs that would have allowed for high-density housing that would have lessened the displacement and in the Pluit Reservoir brought to Jakarta waterfront a model of "socially attuned housing" that could make the city's river normalization a triumph rather than contestation between citizens and the state.[38] Contestation rather than collaboration was what happened in Pluit, in Kampung Pulo and Bukit Duri, and in virtually every case of "restoring the rivers" and continued a pattern of interaction between government and its marginalized residents that had deep historic roots. Goh notes that

> the environmental challenges and displacement threats that *kampung* residents face are an outcome of historical determinants – the segregation of functions and ethnicities during colonial times, the relegation of poorer residents to riskier waters-edge and lowland locations in the post-independence years – and the present-day real-estate development pressures.[39]

Back in 1973, a World Bank appraisal conducted in conjunction with an urban development project in Jakarta noted that nearly 80% of the population lived in unimproved or marginally improved kampungs lacking access to proper drainage and city services and were heavily situated along the city system of rivers and canals.[40] It was also in 1974 that Indonesia established a national housing authority which for the next two decades largely built housing in Jakarta that was inaccessible to the poor. The poor continued to congregate in larger numbers in these vulnerable, flood-afflicted areas with the unofficial sanction of the government. It is not unreasonable to suggest that "normalizing" the rivers was more than flood mitigation. It was a way to remove the vestiges of previous policy failures, casting the onus of Jakarta's flowing sewers on those whose only fault was not having access to the services that citizenship in the city should have afforded them.

Notes

1 Winayanti, L. and Lang, H. (2004) "Provision of Services in an Informal Settlement," *Habitat International*, 28 (1): 42.
2 Ibid., p. 44.
3 Buyamin, A. and Kartini (1998) "Forced Evictions in Jakarta," in Fernandes, Kenneth, ed. *Forced Evictions and Housing Right Abuses in Asia, 2nd Report 1996–97.* Karachi: Eviction Watch Asia, City Press.
4 *Jakarta Post*, January 13, 2010; Eerd, Maartje van and Banerjee, Baryashrea, eds. (2013) *Eviction, Acquisition, Expropriation and Compensation Practices and Select Case Studies: Working Paper 1.* Nairobi, Kenya: United Nations Human Settlements Program.
5 Winayanti and Lang, pp. 55–60.
6 Leitner, Helga and Sheppard, Eric (2017) "From Kampungs to Condos? Contested Accumulation Through Displacement in Jakarta," *Environment and Planning A: Economy and Space*, 50 (2): 445.
7 Kurniawati, Dewi (2009) "The Floods: A Swelling City Is at the Root of the Problem," *Jakarta Globe*, July 24, 2009.
8 *Jakarta Globe*, July 24, 2009.
9 World Bank (2016) "Keeping Indonesia's Capital Safer From Floods." Accessed February 24, 2019: www.worldbank.org/en/news/features/2016/01/08/keeping-indonesias-capital-safe-from-floods.
10 World Bank (2017) *Jakarta Urgent Flood Mitigation Project Overview*, p. 2. Accessed August 12, 2020: www.worldbank.org/P111034/jakarta-urgent-flood-mitigation-project/en/overview.
11 Ibid., pp. 2–4.
12 Wilson, Ian (2016) "Out of the Rubble: Jakarta's Poor and Displaced Seek a Vehicle for Their Voice," *Indonesia at Melbourne*, October 4. http://indonesiaatmelbourne.unimelb.edu.au/out-of-the-rubble-jakartas-poor-and-displaced-seek-a-vehicle-for-their-voice/.
13 *Jakarta Globe*, June 16, 2013.
14 Disaster Relief Emergency Fund, *DREF Final Report: Indonesia's Floods, June 28.* Accessed December 28, 2018: http://reliefweb.int/sites/reliefweb.int/files/resources/Indonesia%20Floods%20-%20DREF%20MDRID007%20Final%20Report.pdfhttp://reliefweb.int/sites/reliefweb.int/files/resources/Indonesia%20Floods%20-%20DREF%20MDRID007%20Final%20Report.pdf.
15 Indonesia Red Cross, www.ifrc.org.

16 Dovey, Kim, Cook, Brian, and Achmadi, Amanda (2019) "Contested Riverscapes in Jakarta: Flooding, Forced Eviction and Urban Image," *Space and Polity*. Accessed October 16, 2019: https:doi.org.10.1080/13562576.2019.1667764.
17 Wilson, op. cit.
18 Leitner and Sheppard, op. cit.
19 Hellman, Jorgen and Voorst, Roanne van (2018) "Claiming Space in Jakarta: Megaprojects, City Planning and Incrementalism," in Hellman, Jorgen, Thynell, Marie, and Voost, Roanne van, eds. *Jakarta: Claiming Spaces and Rights in the City*. London: Routledge, pp. 166–168.
20 Gunawan, Akmad, Winarso, Haryo, and Argo, Teti Armiati (2018) "Impact of Relocation on Livelihood Change of Lower-Income Community. Case Study: Relocation of Kampung Pulo to Rusunawa Jatinegara Barat, DKI Jakarta," paper presented at Planocosmo 4th International Conference, Institute of Technology, Bandung, April 3.
21 Wilson, Ian (2015) "The Politics of Flood Alleviation in Jakarta," *Jakarta Post*, September 5, p. 5.
22 Dovey et al., pp. 10–11.
23 Renzi, A. (2018) "Jakarta: Social and Housing Justice Should Not Be a Gamble on Global Market's Table," *The Conversation*, January 30; Sofian, E. (2018) "Jakarta's Urban Poor Have Found a Way to Fight City Hall," *The Guardian*, April 4.
24 Munk, David (2016) "Jakarta's Eco Future? River Community Goes Green to Fight Eviction Threat," *The Guardian*, November 24; Arbi, Ivany Atina (2017) "River Regions Forces Jakartans to Cook Outdoors," *Jakarta Post*, January 16.
25 Lee, Zachary (2016) "Lost, Jobless: Life After Relocation From Jakarta's Fish Market," *Rappler*, April 18. Accessed March 26, 2020: http://rappler.com/world/regions/asia-pacific/Indonesia/bahasa/englishedition/129663-relocation-pasar-ikan-jakarta-status.
26 Kusumawijaya (2016) "Jakarta at 30 Million: My City Is Choking and Sinking – It Needs a New Plan B," *The Guardian*, November 22; Mariani, Evi (2016) "Get Off the Square! The Unsubtle Gentrification of Jakarta's Old Town," *The Guardian*, November 22.
27 Arbi, op. cit.
28 Sofian, op. cit.
29 "Some 5,000 East Jakartans Affected by Floods," *Jakarta Post*, February 5, 2018.
30 Thynell, Marie (2018). "Urban Inequalities in a Fragile Global City: The Case of Jakarta," in Hellman, Jorge, Thynell, Marie, and Voorst, Roanne van, eds. *Jakarta: Claiming Spaces and Rights in the City*. London: Routledge, p. 31.
31 Wilson (2015), op. cit.
32 Putri, Parthiwi Widyatmi (2019) "Sanitizing Jakarta: Decolonizing Planning and Kampung Imaginary," *Planning Perspectives*, 34 (5): 818.
33 Bakker, Karen (2007) "Trickle Down? Private Sector Participation and the Pro-Poor Water Supply Debate in Jakarta, Indonesia," *Geoforum*, 5: 860.
34 Ibid., pp. 860–861.
35 Ibid., pp. 862–863.
36 Dovey et al., op. cit.

37 Goh, Kian (2019) "Urban Waterscapes: The Hydro-Politics of Flooding in a Sinking City," *International Journal of Urban and Regional Research*, 43 (2): 250–272. doi: 10.1111/1468-2427.12756.
38 Ibid., p. 265.
39 Ibid.
40 World Bank (1974) Appraisal of the Jakarta Urban Development Project in Indonesia, August 30, Report No. 475 IND. Washington, DC: World Bank.

7 Jakarta's present and future of flood risk management

January 2020 marked the beginning of the eighth decade of the Republic of Indonesia and Jakarta's role as the nation's capital. Flood mitigation, the provision of adequate drinking water; the proper disposal of waste; eliminating pollution of the rivers, streams, and canals, and addressing the problems of a sinking city remained unresolved challenges in the day-to-day management as Jakarta entered 2020. Despite at least 30 years claiming that these were all priorities of the government, the New Year dawned, like so many annual beginnings over this span, with vast areas of the metropolis under water. Early morning on January 1, the rains began, dumping approximately 400 millimeters (15 inches) in a few hours and continuing until January 2. This prompted the usual mass evacuations to higher ground of residents located along the cresting rivers, with some communities registering up to four meters of water when the nearby river peaked.[1]

Flooding persisted for a week, crippling large areas of the city and surrounding metropolitan jurisdictions. As the waters receded, recriminations rang out from those affected. Affected businesses filed a lawsuit against the Jakarta government claiming their losses were due to ineffective flood management. Anies Baswedan, then in the third year of his governorship, took the brunt of the criticism that included protest rallies in front of City Hall demanding his resignation. According to the class action lawsuit filed in the Central Jakarta District Court by the Indonesian Shopping Centers Tenants Association, their members lost 43.32 billion rupiah (US$3.17 million) because the flood closed 300 shops.[2] All together, the January 2020 flood forced evacuations of over 500,000 persons and accounted for 60 deaths. The extensive interventions to manage the heavy rainfall, including the near completion of the river dredging project (Jakarta Emergency Dredging Initiative [JEDI]), clearly had not mitigated all the flood problems in Jakarta. The JEDI project did make a difference in some routinely inundated areas. According to a Jakarta government official, flooding had not been so bad in North Jakarta, an area of the city usually hard hit during flood events.[3] Actually, the worst flooding occurred in peripheral areas just outside the city to the east and west, places previously not prone to flooding. Although much of this was outside Jakarta's jurisdiction, this did not stop the victims directing their criticism at Governor Baswedan, contending his administration was too slow in reacting to the elevated water in the

DOI: 10.4324/9781003171324-8

rivers that swamped their communities. Whatever actually caused the mass flooding in the first week of January 2020, it was clear that Jakarta's flood risk management system was not working. Jakarta's communities would continue to experience to extreme flooding especially during the three to four months of the rainy season. After 30 years of efforts to mitigate floods, Jakarta seemed little better able to manage its waters than it had been in the 1990s.

The post-flood postmortem echoed a familiar theme, namely that record rainfall was largely to blame for the January 2020 flood. Apart from the lawsuit against the Jakarta government the local media coverage did not contain a serious critique of the mitigation efforts that had been instituted and certainly not the in-depth assessments that followed the 2007 flood. Why? One explanation might be that fixing the flooding problem in Jakarta was old news, and the real news was about new challenges to the primacy of the capital city within the region. The 2020 flood struck Jakarta less than six months after President Widodo, just reelected to a second five-year term, announced that Jakarta was going to be replaced as the capital city. Over the previous five years, the Indonesian government had been working on a plan to move the capital city away from the disaster-plagued land on which the megacity sat. In late April 2019, two weeks after completion of the national election, Joko announced that the government had been studying potential sites for constructing a new capital city. No specific location was announced at the time, just that it would be somewhere outside of Java. The site designated needed to be free of the variety of problems that plagued Jakarta, including flooding, land subsidence, pollution, traffic congestion, tsunamis, earthquakes, and forest and land fires. Another criterion was that it should be more centrally located within the Indonesian archipelago, which was why no sites on Java worked. This made provinces on two islands, Kalimantan and Sulawesi, the most likely candidates for the new capital. Both were the more centrally located than Java or any other islands, although both could be faulted on several of criteria related to the low probability of being affected by Indonesia's variety of natural and man-made disasters. Sulawesi was not immune to tsunamis and for years Kalimantan generated severe air pollution because its vast agricultural sector relied on the practice of slash and burn to remove vegetation. Although no specific sites were announced, speculation initially focused on Central Kalimantan and its capital city of Palangkaraya. This was where President Sukarno planned to relocate the capital away from Jakarta in the 1950s. Sukarno, the consummate infrastructure-builder president, admired and sought to replicate in Indonesia the plan for Brazil's new capital city, Brasilia. Although the government completed a master plan for the new capital, the project failed to progress beyond some partial implementation. The current street plan includes a small section of Palangkaraya with the formal, wide boulevards that Sukarno admired in the Brasilia scheme. Otherwise, this remote area really had little to suggest it work as a future capital city.[4]

Jokowi actually hinted at the need to move the capital to a new location to foster more dispersed national development during his campaign for reelection. In campaign speeches, he stressed a commitment shift development of the nation beyond the current Java-centric pattern that had been inherited from the New

Order regime. This played extremely well in Kalimantan where he handily carried four of its five provinces in the election. What was not revealed until after the election was that the initial background study considered three options. One was to move all of the governmental operations into a single district surrounding the presidential palace and the Monas in Central Jakarta, which would make a capital district. A second option involved moving the capital to a site within reach of Jakarta but separated from the high density of the megacity that hindered the efficiency of the national government. According to National Development Planning Board Minister Bambang Brodjonegoro, who headed the study, "neither of those two options would address the overpopulation in Java, a home to 57 percent of the roughly 260 million people of Indonesia." Nor would they "support the government's aim to shift the nation from its Java-centric development to a more inclusive development agenda for the whole archipelago."[5] So off Java was where it would go. No matter what location the government chose, it required a substantial upfront investment to create a new capital city. According to estimates prepared by the study team, costs for a new capital city could run to US$30 billion depending upon how much of the government relocated there. During the initial days following announcement of the proposed new capital city, reactions in the media tended to be quite positive. Many picked up on the theme that Jakarta was a megacity in deep trouble, with seemingly unmanageable problems like regular and devastating floods and environmental degradation. Later in 2019, the government announced that East Kalimantan had been selected as the site for the new capital city utilizing land that the government owned (thereby helping to reduce some of the development costs) located between the cities of Samarinda and Balikpapan.

Almost on cue, nature stepped in to demonstrate how it could disrupt life in Jakarta and to affirm the merits of relocating the capital. Just prior to the president's announcement in 2019, Jakarta was hit by a highly unusual and intensive April rainfall with flood conditions inundating approximately 37 Jakarta neighborhoods, mostly in East and South Jakarta, when the Ciliwung, Angke, and Krukut rivers crested because of unusually heavy flows coming down from Bogor. This was Governor Baswedan's first major confrontation with flooding and he promptly announced plans to jump-start what he termed river "naturalization" projects. The switch in nomenclature from "normalization" to "naturalization" was deliberate. While normalization was associated with removing kampungs along the rivers and canals, a policy Baswedan stated during the 2017 election campaign that he did not support, "naturalization" emphasized efforts to improve drainage (though construction of infiltration wells) and handling increased river flows through more retention ponds and reservoirs. He also supported improvements to the seawall in North Jakarta to protect rather than remove riverfront informal settlements. These "naturalization" efforts, under the jurisdiction of the Jakarta Water Resources Agency, also included greening a 500-meter section of the Ciliwung on Jalan Krapu in North Jakarta and a 600-meter section of the West Flood Canal in Dukuh Atas. The Sunter, Kampung Rambutan, and Mimangis reservoirs would be naturalized by transforming riverbank areas into public spaces which would likely cause some displacements. The national government's normalization program under the Ciliwung–Cisadane Flood Control Office (based in the Ministry of Public Works and Housing) planned to continue "normalization," but Baswedan's "reluctance to carry out evictions since he took office in 2017" stymied its progress.[6] The Baswedan government also looked beyond its borders for ways to mitigate flooding. It began discussion with the Bogor city administration to add retention areas and several dry dams upstream to regulate the flow downstream. These efforts demonstrated a sustained commitment to flood mitigation, although the emphasis on incremental interventions reinforced the impression that there was no grand plan to address the water problems of this "sinking city." In fact, there was a grand strategy in the works, but not one to correct Jakarta's problems immediately.

In July 2019 and on the heels of the announced decision to move the capital city off Java, the national government unveiled a revised version of the controversial National Capital Integrated Coastal Development initiative to address Jakarta's recurring flooding problem. The original "Great Garuda" plan (discussed in Chapter 5) involved the construction of a series of offshore islands by private developers. The "Great Garuda" project started in 2012 with the construction of a few islands and then halted because of administrative wrangling over the authority to carry it out. The new plan, as shown in Figure 7.1, called for the construction of a massive dike through land reclamation, mirroring the strategy successfully employed in the Netherlands. The two-part sea dike, that would take 30 years to complete, would begin on the shores of Cengkareng near Jakarta's international airport, extend into the sea across North Jakarta, and

Figure 7.1 Jakarta Coastal Defense System project rendering.

Source: Giant Seawall Jakarta National Capital Integrated Coastal Development, Jakarta, 2017.

terminate on shore near the suburban city of Bekasi in the east. The completed structure would stretch 32 kilometers from Cengkareng to Bekasi. It called for a toll road constructed on its surface providing access to Islet G, a reclamation space created when the original waterfront development initiative was still in play. Another 2,000 hectares of reclaimed land for urban development would help to underwrite a portion of the 262.2 trillion rupiah (US$18.7 billion) cost of the project. Unlike the plan to move the capital to an off-Java site that was a done deal, the announcement of the sea dike project seemed more like a trial balloon. Already Indonesia had executed agreements with the Dutch and South Korea governments to support the project. But the project was not scheduled to begin until 2030 and long after the current government was a distant memory.[7] As the *Jakarta Post* editorial appropriately observed, this new plan included no mention of efforts to address the current problems, to stop the sinking of Jakarta by restricting groundwater extraction, to improve the capacity for aquifer recharge, or to develop alternative surface water sources or address surface water pollution. The sea dike plan discounted the concerns expressed since the mid-1990s by environmentalists, maritime scholars, and fishing communities that land reclamation along the waterfront would create new environmental, economic, and social problems.

The sea dike scheme announced in 2019 revived discussions about the value of continued construction of the seawall lining the coast that had been underway in a piecemeal fashion since 2014. The Ciliwung–Cisadane Flood Control office managed construction of this shoreline seawall that was a direct responsive to the flood in 2013 that had hit so hard in the upscale north coast communities in Pluit, including Pantai Mutiara. In the view of Tuty Kusumawati, head of Jakarta's regional development planning board, "it is urgent that the seawall be built, otherwise Jakarta will drown."[8] As the construction proceeded in 2016 under the auspices of the national government, the Jakarta government observed that the seawall alone "would not function optimally if groundwater usage was not limited."[9] Later that year, construction halted as the national government reexamined the project to assess potential negative environmental impacts. By early 2017, construction resumed, and the Minister of Public Works and Housing announced completion of the first 4.5 kilometers of the seawall protecting Muara Kamal, Pluit, and Muara Baru. The rest of the first 20 kilometers would be done by the year's end. Yet by late in 2018, only 90% of the seawall construction had been built. The remaining sections along Pasar Ikan and Kali Blencong would be completed in 2019. Actually, the initial work in Muara Kamal announced in 2017 actually was not undertaken because residents there did not support the government plan. The newly elected Governor Baswedan had honored his commitment to Muara Kamal to take a gentler approach to settlement removal when undertaking measures in the name of flood mitigation. More than likely the Muara Kamal residents' opposition arose as they witnessed the complete demolition of nearby Pasar Ikan to facilitate seawall construction.[10]

Other major infrastructure initiatives in Jakarta proceeded on schedule. Early in April 2019, prior to the announcement of the plan to move the capital from

Jakarta, and prior to the flash flood that followed, Jakarta officially opened for business the long-awaited Mass Rapid Transit system. The city's first transit line ran underground from the Hotel Indonesia business district south to Senayan and then above ground over Jalan Fatmawati to Lebak Bulus just south of the luxurious Pondok Indah community in South Jakarta. This served one of the busiest and most congested gateways connecting South Jakarta to the city's center. Jakarta's quest for a mass transit system had been an on-and-off infrastructure dream since the 1970s. According to Ali Sadikin, during his governorship in the 1970s, he proposed the construction of a subway along a route similar to the one taken by the MRT, but the National Development Planning Board said it had no money for it.[11] It went back on the drawing board in the 1990s as an infrastructure necessity for what had become a "mega" capital city. Serious planning of the MRT resumed in 1996 and included a funding scheme. It was linked to the waterfront project since the proposed subway terminus would be in the old town (Kota) that was the hub of the revitalization scheme. Bimantara, the business firm of Suharto's second son, Bambang Trihatmodjo, was among the project investors. So sure was the belief that the mass transit system was to become a reality that in March 1997, the *Jakarta Post* announced in bold type on its front page that the "Subway Will Start in June." A Japanese-European consortium had the green light from the government for what was calculated to be a US$2.2 billion undertaking.[12] Within months, however, the bottom began to collapse on the Asian economy, especially in Indonesia. With the onset of a full-blown financial crisis by the end of 1997, planning for the subway project ceased. Although the Bus Rapid Transit system that began under Sutiyoso's administration several years later partially filled the city's enormous public transit gap, the need for a rail transit link between the center and the periphery continued. During the Jokowi governorship in 2013, construction of the MRT commenced for real. The MRT opened in April 2019 (see Figure 7.2) and offered a speedy alternative to the routine congestion along the major north-south Jalan Fatmawati–Jalan Sudirman corridor. Passengers could connect with Jakarta's other public transport services, including the Transjakarta busway system, the commuter railroad that intersected at the Dukuh Atas stop, and a light rail system still under construction intended to serve outlying districts. The first phase of the MRT was still in its maiden voyage when work began to continue the line underground from the Hotel Indonesia to Kota in North Jakarta. Then would come an 87-kilometer east-west line as shown in Figure 7.3.

To support ridership and to manage the use of surface traffic, Jakarta's transit company used its right-of-way to promote transit-oriented business and residential development. These developments offered a way for the system to generate operating revenue by selling or leasing lands it controlled. This allowed MRT fares to remain low enough to promote ridership. Simultaneously, three other companies were building the light rail system to extend high-speed transit service to communities south and east of Jakarta where the greatest population growth in the metropolitan area had occurred since 2000. Concurrently, the national government contracted with a Chinese firm to construct a high-speed train link between Jakarta and Bandung.[13]

Figure 7.2 Passengers in the new Jakarta MRT in April 2019.
Source: Photo by author.

In light of these substantial investments in transportation infrastructure, coupled with the commitment to build a new capital city, what was the immediate plan to address the deficiencies in Jakarta water management system? What about the polluted rivers, and the less than desirable quality of and access to clean water for Jakarta residents? What about the land subsidence problem and the unsustainable reliance on groundwater to meet the city's needs? What about regreening in light of the potential impact of even more intense land development spurred

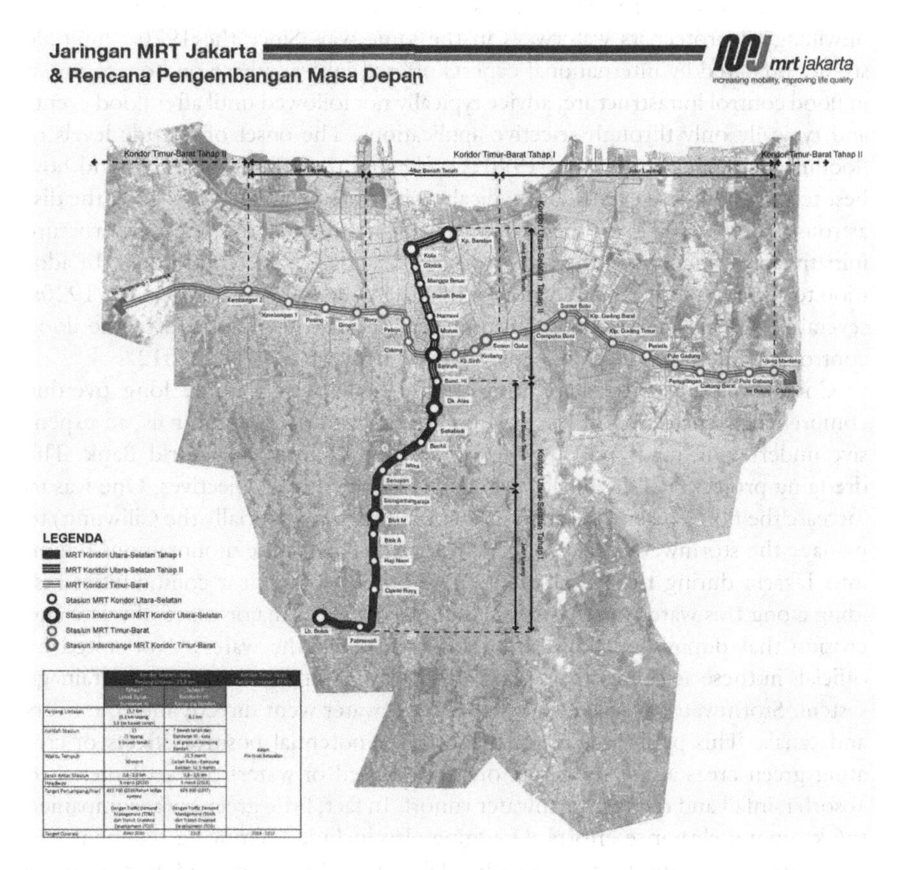

Figure 7.3 Map showing the current and future MRT lines in Jakarta.
Source: MRT Jakarta, Jakarta.

on by transit development into the periphery? Moreover, what about the governance structure needed to manage rivers whose floodwaters crossed multiple jurisdictions before inundating Jakarta communities and businesses? Was the current game plan of piecemeal interventions sufficient to prevent Jakarta from becoming, as some predicted, completely submerged by 2050?

It is worth revisiting key components of Jakarta's overall flood management system as it evolved over the past few decades to best gauge present conditions and to envision a possible path to a different future. Since the 1970s, and evidenced by the recent decision to create several additional physical flood mitigation barriers, like the seawall across the north coast shoreline, and potentially a sea dike further out in Jakarta Bay, Jakarta adhered to a strategy similar to what the Dutch employed to manage water in colonial Batavia, namely engineered infrastructure interventions to block high waters. Unlike the current practice in the Netherlands that is largely anticipatory rather than reactive, and that safeguards the waterways from the effects of urbanization, Jakarta's approach has been reactionary and unwilling to protect its waterways in the same way. Since the 1970s, multiple studies prepared by international experts offered Jakarta advice on how to invest in flood control infrastructure, advice typically not followed until after flood events and typically only through selective applications. The onset of historic levels of flooding beginning in the mid-1990s accelerated the efforts to understand how best to mitigate these events, but typically with little or no action. It took the disastrous floods of 2002 and then 2007 to instigate the long overdue infrastructure initiatives as evidenced in the rapid completion of the East Flood Canal. In addition to completing this final section of the flood canal system begun in the 1920s, several new drains and addition of pumping capacities were added to the flood control system under the Bowo governorship between 2007 and 2012.

Complementing these hard infrastructure additions was the long overdue comprehensive dredging of the existing channelized rivers and canals, an expensive undertaking made possible with loan support from the World Bank. The dredging project (JEDI), finally begun in 2013, had two objectives. One was to increase the flow capacity of Jakarta's main waterways (especially the Ciliwung) to manage the stormwaters that regularly race down from the mountainous region into Jakarta during the rainy season. The other was to clear communities residing along this waterway that officials claimed to be a major cause of waste and erosion that diminished flow capacity and polluted the waters. Not noted by officials in these assessments was Jakarta's lack of a separate stormwater drainage system. Stormwater as well as household gray water went directly into the rivers and canals. This practice counterbalanced any potential positive effects of creating green areas along the rivers on land cleared of waterfront settlements to absorb rainfall and reduce stormwater runoff. In fact, little greenery accompanied the kampung clearance efforts. Dredging also included removing sediment and debris from lakes and reservoirs within Jakarta (and the surrounding region) so that they could better handle excess water during flood conditions. Yet, as critics pointed out, rebuilding the channel walls along the rivers, which was part of the JEDI strategy, increased the intensity of the floodwaters passing through Jakarta communities and would not necessarily mitigate the problem.

Designers of the JEDI project acknowledged the extensive social costs for thousands displaced by the process of river "normalization." Some displaced families secured upgraded housing, but this often meant moving to more expensive accommodations or places too far from where they earned their living. As a result, displaced residents often turned down the replacement housing in lieu of moving into another nearby informal settlement not yet confronted with demolition. Although it was possible to argue that these informal settlements contributed to the degradation of the rivers, officials never seriously studied how much they contributed to the problem. The settlements were illegal, too close to the rivers and canals, they obstructed the dredging, and keeping them was not an option. Were these communities, if fact, the main contributors to the pollution problems of Jakarta's rivers? Images in the media of massive piles of cut logs blocking the drains in Jakarta flood control system during the heavy rains were just one indication that upstream activities contributed to much of the debris and likely the pollution that fouled the rivers. Numerous studies of the sources of pollution and debris in the rivers in Jakarta point to the significant impact of upstream activities. When calculating the social costs of the dredging efforts borne by displaced communities, coupled with the addition of new hard infrastructure along the embankments, did this yield the desired improvements to Jakarta's rivers and an effective flood mitigation strategy?

Offsetting any potential benefits of increasing the flow capacity of Jakarta's rivers by dredging to reduce flooding impacts and river "normalization" is the now firmly documented condition of Jakarta as one of the fastest sinking cities in the world, a condition that exacerbates the flood impacts and evidence of its problematic water management system. With between 40% and 50% of the city's land situated under the mean sea level, the now expanded flow capacity of Jakarta's dredged rivers has nowhere to go when it reaches the north coast unless pumped over the seawall. Otherwise, it overflows into the adjacent communities. Expansion of pumping capacity at the seawall along the north coast manages the river waters that no longer flow unassisted into the sea. As Jakarta's land continued to sink, however, this required not only more pump capacity but also continuous additions to the top of the wall in order to keep the Java Sea from spilling over into the city.

The consensus view on the causes of Jakarta's land subsidence focus on two factors. One is that development pressures contribute to soil compaction and that accounts, in part, for the sinking process. The other factor, unregulated reliance on groundwater extraction causing the collapsing aquifer, is the main culprit, however. Evidence of land subsidence first appeared in the 1990s and prompted studies that measured the extent of the problem. The study by Abidin et al. (2011) verified the connection between the clean water needs of the rapidly growing megacity and land subsidence.[14] As the aquifer compressed, saltwater intrusion into groundwater system was an additional problem since it contaminated wells in North Jakarta communities. Excessive groundwater extraction was necessary since surface waters remained unfit for human or even industrial consumption. Even the treated surface water is inferior. So private wells, many puncturing into the deep aquifer, extracted massive volumes of water to serve the city and without any capacity for recharge. The subsequent land subsidence affected the flow of the rivers and removed the natural process of self-flushing to remove pollutants. This also required greater reliance on pumping stations, like the one on the north end of the Pluit Reservoir along the coast, to remove potential floodwaters.

So if land subsidence is such a critical factor affecting Jakarta's persistent flooding problem, why has there been so little effort to address it? One explanation is that land subsidence occurs "gradually and invisibly," unlike flood events that can quickly galvanize political action. As Colven (2000) notes, "managing and regulating groundwater extraction and land subsidence is largely work with 'very little public relations value.'" At the same time, flood control projects are "closely aligned with an ingrained world-class city aesthetic, making them highly attractive to politicians with ambitious urban development agendas."[15] The great seawall project represented the most "visible" response to flooding, but a response that did not address the core problem. On a day-to-day basis, river and canal dredging, removing informal riverfront settlements, and trash removal also garnered some political and public attention. At the same time, public concerns over new developments that violate plans to protect green areas for rainwater absorption and aquifer recharge dissipate quickly once the sunshine returns.[16] Caljouw, Nas, and Pratiwo observed in the aftermath of the 2002 flood that there was widespread criticism of government's response. It targeted Governor Sutiyoso as the city's chief executive responsible for flood control that did not prove effective. It also singled out developer Ciputra whose high-end residential enclaves were unaffected by the flooding that inundated less affluent and less protected neighborhoods. President Megawati got a share of the public rebuke for what many saw as a slow response by the national government. Although the immediate aftermath of the 2002 flood generated such highly charged political fallout and widespread calls for better government action, the clamoring faded relatively quickly. Caljouw, Nas, and Pratiwo discovered when they visited the city several years later that

> the people of Jakarta give the impression of having forgotten the tragedy ... People have stopped talking about it. The garbage that was blamed as one of the factors causing flooding has now ceased to be a topic of debate, and garbage is still floating everywhere in the rivers and canals of the city ... There is no more public debate about the problem of water-catchment areas.[17]

So even the visible disasters lack staying power when there are no concrete initiatives generated by the disaster except for diffuse comments that "something" needed to be done. Shortly after these researchers published these observations, the historic 2007 flood hit with even more devastation, and this time, political fallout had more staying power.

As previously discussed (Chapter 5), the 2007 flood brought the issue of land subsidence front and center to explain the extent of the devastation. Jakarta was now described in the local and international press as a sinking city. So land subsidence was highly visible and widely discussed. The lack of political urgency to address land subsidence is best explained by the high cost of addressing this "underground" problem. Based upon expert assessments, the only credible way to reduce the demands on groundwater use is to substitute surface water. The investment necessary to clean up and keep clean the surface waters includes the construction of sanitary infrastructure to prevent continued pollution, the addition of new water retention spaces to augment the supply of clean surface water, and water treatment facilities. Implementing surface water management throughout the watershed, removing stormwater drainage directly into the rivers and canals, and extending wastewater service to communities currently using the rivers require extraordinary expenditures that government never has been willing to make. The microscale of the river clean-up is best illustrated in the transformation of a small segment of the Kali Besar situated between Sunda Kelapa on

Figure 7.4 Pollution in Kali Besar in the Kota area.
Source: Photo by author.

the north coast and historic Fatahillah Square. Five years ago, the pollution (as shown in Figure 7.3) was pervasive. A return to the same spot two years later shows the same water visibly much clearer (and perhaps less polluted) as shown in Figure 7.4. This also coincided with the separation of the upper part of the kali to create a pedestrian space within a clearer and less smelly waterbody (as shown in Figure 7.5) to support increased public access to this historic space.

The problem with deferring actions on system upgrades is that the incremental interventions, such as evidenced in the cleanup of Kali Besar, cannot address the underlying problems. In the Pluit Reservoir located in North Jakarta, as in all of Jakarta's rivers, the daily dredging program underway is necessary to keep pace with the accumulation of solid waste that blocks the drains, adds new sedimentation and pollutants, and increases the likelihood of flooding during the heavy rains. Drainage pump failures happen often, not only because of the total clogging due to the accumulation of debris during a heavy storm but also because of damage caused when the pump propellers are hit by hard materials.

Of course, solid waste problem is not just from Jakarta communities since a vast quantity of it comes from upstream sources. Stormwater frequently brings forestry and agricultural debris into the rivers to add to the urban solid waste burden. This is difficult to prevent because none of the rivers within the

Figure 7.5 Kali Besar waters cleaned up to encourage visitation to Kota area.
Source: Photo by author.

Jakarta watershed benefit from a single unified system of management. Rivers in Indonesia are under the legal authority of the national government but with daily management responsibilities shared between many units of local and provincial governments. Groundwater is under the jurisdiction of the national ministry that overseas mining, but it is up to the local government to institute measures to regulate its extraction. As a result, there is no one authority responsible to address these cross-jurisdictional challenges of groundwater and surface water management. As Bakker (2007) succinctly noted: "Jurisdictional fragmentation has reduced the ability of any one level of government in Jakarta to effectively govern water resources within a watershed, or even within urban boundaries."[18]

The unmanaged (and seemingly unmanageable) stormwater, solid waste, and wastewater problems in Jakarta's rivers reflect the lack of regulation over the location of building construction within the watershed. Jakarta's robust tradition of preparation of plans and planning regulations since the 1950s lacks a complementary tradition of robust enforcement to achieve compliance with the plans. Governmental absence from enforcement contributes to the loss of open space, the failure to prevent new development in flood-prone areas, and encroachment of urbanization on waterways. In addition, limited authority to require proper water and sanitary infrastructure in new formal and informal development, and a

Figure 7.6 Kali Besar redesigned to encourage visits to the Kota area.
Source: Photo by author.

lack of enforcement of laws against point source pollution emanating from residential, commercial, and industrial uses perpetuate the polluted conditions. The shortcoming in Jakarta's planning implementation processes evoke continuous criticism from citizens now empowered to hold the government accountable. But the fact that one can stroll down the walkways of virtually any "traditional" riverfront community and see pipes from the houses sticking out over the waterways to dispose of its wastewater indicates that the problems and deficiencies are obvious and obviously ignored.

In a provocative assessment of the deficiencies in Jakarta's approach to flood mitigation, civil engineer Sedlar notes that the flooding frequently occurs where it should not have occurred based upon the infrastructure investment strategies undertaken. He contends that flooding is less about the intense precipitation and overflowing canals, but rather emanates from various anthropogenic causes. Sedlar points to "deforestation upstream" as an explanation for the walls of water rushing into the delta region during the rainy season. The inadequacy of Jakarta's pumps to remove high waters from the urban waterways that have sunk below sea level (and cannot drain themselves naturally into the Java Sea) and the malfunctioning and obsolete floodgates that triggered flooding where it should not occur based on topography are also tied to anthropogenic factors.

Finally, he agrees with others who contend that the seawall along the North Jakarta coast cannot keep up with the combination of sea-level rise in the Java Sea and the concurrent land subsidence affecting the wall itself on the landside. He maintains that modeling future flood problems and responses based upon historic patterns has not worked to identify the places and timing of floods. Changes in the city's development have caused flooding in places the infrastructure cannot protect. One alternative is to access real-time data through social media to provide for quicker and more appropriate responses to emerging flood conditions. With these data, it is possible to spontaneously map places where flooding is occurring so that infrastructure responses adjust to the changing circumstances.[19] This citizen-engaged participatory method of analysis corresponds to the unpredictability of flooding in a sinking city with flood control infrastructure created for a different reality. Real-time mapping might also make it possible for a deeper and more "visible" understanding of flood factors such as land subsidence and bring it more directly into the urban policy discourse.

The transition in Indonesia to a governance system that legitimized bottom-up planning following the decentralization legislation beginning in the late 1990s placed greater responsibility for flood management in the hands of the local government as it relates to urban development and protection of natural resources, especially the urban rivers and other surface water resources. The tradition of planning in Jakarta, however, offers little evidence of considering impacts on natural resources when formulating development schemes. From the city's beginnings, allowing development to push right up to the city's waterways never raised any alarm bells. It is possible to find references in recent Jakarta's planning documents to objectives calling for protection of environmental conditions but without reference to any specific strategies to realize these objectives. Currently, there is a national "objective" that 30% of the city's land must be kept as green open space. As Manan (2016) shows, the current national mandate is consistent with the objectives set forth in the city's 1965–1985 master plan that set the open space target at 37%. Lack of management and the predominance of market forces explain why currently Jakarta is nowhere near that standard. Less than 10% of the Jakarta's land area, including the undeveloped islands in the Java Sea, remains as green space.[20]

Not until the administration of Ali Sadikin in the late 1960s and 1970s was there any regulation of settlements along the city's rivers. This also coincided with the initiation of his highly acclaimed Kampung Improvement Program to upgrade these unserviced neighborhoods by adding some environmental infrastructure. Sadikin's interventions did not include greatly expanded access to clean water or connections to sanitary waste systems in most cases, although there were improved footpaths, better drainage, and some communal toilet facilities that benefitted some of the kampungs. The government's approach confirmed that low-cost communities, unlike more prosperous neighborhoods, did not warrant the water services of the modern city. Abeyasekere notes that by the end of the Sadikin era, only 18% of Jakarta had access to piped water, and few of the connections resided in the kampungs. According to a World Bank report, only between 15% and 20% of the municipal budget was spent in kampungs even though these residences accounted for 60% of the population. As Abeyasekere put it in assessing the response of the Sadikin government, "the Jakarta government chose the cheapest means of assisting the poor, so as not to jeopardize their concentration on building up the more visible parts of the city as a modern international capital."[21]

Under Jakarta's governor Sutiyoso, who presided over the first decade of decentralization during Indonesia's economic crisis, and dealt with Jakarta's historic floods in 2002 and 2007 (as discussed in Chapter 5), the idea of "normalizing" the riverfronts on a massive scale began. Even though government officials and the donor agencies that financed normalization expressed concern for the social and economic costs of flood management, most assessments of how to mitigate flooding regarded it essential to enforce the prohibition of settlements on the riverbanks. Sutiyoso's administration took a tough stand with the riverbank residents that generated contestations between residents and the government even when replacement housing, often of a higher quality (but also often at a higher cost), was available.

Flood mitigation, not social improvements, was the rationale for removing the informal settlements. No consideration was given to the alternative to bringing urban services, such as clean water and sanitation, to make these dense, but socially cohesive, communities better places and to enable strict enforcement of laws against polluting. Putri (2019) suggests that the commodification of basic services did not include cheap and reliable water and sanitation series for the communities. At the same time, "bulldozing dilapidated kampungs also incurred the risk of cutting off the flow of inhabitants that sustained the country's low-cost labour market." Even if the water service was provided by the water company, poor communities in Jakarta could not afford the cost of installation because of the connection fees and the monthly carrying costs. Without a "pro-poor" policy on water, there was no way to alter the water habits of these communities.[22]

Fauzi Bowo faced citizen outrage over water management issues during his one term as governor from 2007 to 2012, a spillover from the effects of two major floods under his predecessor Sutiyoso, and ignited again when existing safeguards failed to prevent flooding in 2009. According to Steinberg, the flooding and waste management problems were man-made and fixable by active government intervention. To Bowo's credit, in response to the 2009 flood when capital improvements were not readily available, he instituted an early warning system, and added pumps along the toll road to the airport to forestall flooding on that critical roadway.[23]

It was during Bowo's term as governor that two big projects, the river dredging (which actually hastened "normalization") and the "Great Garuda" waterfront scheme, elevated the heavy infrastructure approach to water management to an even more central strategy. As discussed previously in Chapter 5 Akhmad Hidayatno and fellow systems engineers at the University of Indonesia's Modeling and Simulation Laboratory studied the potential impact of the waterfront redevelopment scenarios advanced over the past several decades to determine whether the proposed Jakarta Coastal Defense Strategy (JCDS) represented the best alternative to advance Jakarta's development interests. As a choice between economic betterments or enhancing environmental qualities or perhaps ideally accomplishing some aspects of both objectives they were inclined to support a compromise strategy involving select new waterfront development with allowances for green open space protected by the seawall would be the most fruitful scenario.[24]

Jakarta's flood problem and the difficulty of finding the right fix was not unique. Across the globe, the recurring problem of flooding in densely populated places heightened awareness of the precarious conditions confronting cities in delta regions and the need to mitigate the flooding through innovative interventions. In Jakarta, managing floods vied with traffic congestion as a measure of the local government's overall capacity to manage the megacity. Yet by examining flooding in the context of overall water management, and doing so through a longer time frame than just the recent past, it becomes clear that the problems required addressing more than just handling the rainfall in the "wet season." Douglass (2010) links the problems of water management to the effects of the transformation of the Jakarta extended metropolitan region (or Jabotabek) into Indonesia's center of "global production and consumption." This intensified "pressure on the urban environment, particularly on the ecology of water systems, ... resulted in the repeated massive flooding" of the city.[25] He notes the failure of planning in the Jakarta region to implement effectively the "guided land development" approach offered by the Dutch consultants (advice provided since the 1970s) "to steer urban growth ... away from the environmentally fragile northern coastline and the southern and southeastern aquifers, hills and high mountains where settlement and deforestation threatened Jakarta with increasing extremes of water shortages and flooding."[26] Accelerated metropolitan development since the mid-1980s negated the potential benefits of Jakarta's regional planning strategy, as it not only eliminated vast sections of the aquifer recharge areas but also the resulting deforestation further clogged the rivers and the increased demands on subsurface water resources, with resulting land subsidence. This contributed to the historically high floods.

The singular focus of analysts and the local leadership on alleviating flooding in Jakarta's inner city areas misses key lessons that become much clearer when considering the full range of water management concerns over the past 400 years. Since the original settlement in the 17th century, there have been three related but distinct challenges central to effectively managing the city's water. One is simply the provision of sufficient water for human and industrial consumption that would seem to be quite easy given the proximity of 13 rivers and their tributaries and a reliable rainy season that sufficiently compensates for any extreme during the dry season. But in fact, supplying usable and reliable water was a problem from early on in Batavia, continued when the urban population expanded into the periphery, and now remains an unfulfilled promise. The majority of Jakarta's 11 million inhabitants are unable to experience clean water in their homes. As discussed in Chapter 4, the extension of government contracts to two foreign companies for exclusive control of Jakarta's water service failed to expand coverage despite being allowed initially to raise tariffs in exchange for expanded service. Privatization did not solve Jakarta's water problems, notes Steinberg, thereby raising the expectations that the public sector can and will respond effectively.[27] Recent actions belie that faith. The Jakarta governor's budget in 2020 underscored the lack of government priority for enhancing the quantity and quality of water when it eliminated a proposed allocation dedicated for enhancing water service. It shifted the funds to support home purchases for moderate and middle-income families. With the Jakarta water company PDAM engaged in a process of resuming full responsibility for the city service when it secures all of the assets of the two foreign providers by 2022, this could potentially expand the level of service. Until then, however, the clean water problems of Jakarta persist. In the meantime, international donor projects, such as the IU WASH PLUS project funded by the US Agency for International Development (USAID), are making progress across urban Indonesia, and to some extent in Jakarta, by assisting kampung residents to connect to water and sanitation services. Nationwide, the USAID efforts enabled more than two million Indonesians to access improved drinking water between 2011 and 2015. In addition, roughly one-quarter million urban residents gained access to improved sanitation. These were the lucky ones, but it represents just a literal "drop in the bucket."[28]

Bringing toilets into urban kampungs to eliminate the need for open defecation is the challenging aspect of the donor efforts at upgrading water and sanitary conditions. Flood control investments divert funding that could be used to provide sanitary infrastructure for a sizeable segment of Jakarta's population. This has been a problem of priorities with a long history. When Jakarta began in 1983 to expanded sewer service, only about 2% of the city had access. Thirty-five years later, according to data from the government agencies, the overall area covered by sewer service was still less than 5%. Despite preparation of several master plans for wastewater management, including schemes to add a centralized sewerage system, implementation proved logistically and financially prohibitive. Public policy continued to ignore the widespread use of the rivers as the primary locus of household and industrial waste for the majority of Jakarta residents. They accepted instead the "small scale interventions led by international organizations" supporting "community-based sanitation systems in informal settlements" such as shared latrines and improved septic tanks.[29] A 2012 master plan proposed a citywide sewerage system by constructing it as 15 separate zones rather than a single centralized system. Yet given the lack of regulation of the built environment above these zones, piping such a system even "zone by zone" was unworkable and with a cost that would be prohibitive.[30]

Ensuring an adequate water supply and dealing with pollution effects of human and non-human impacts on surface and subsurface water are fundamental urban needs sidelined by a policy agenda that made infrastructure investments for flood control the overriding concern. Enhanced sanitary facilities do not fit with urban flood control policy since they remain a "private" responsibility. Yet all three urban needs, clean water, sanitary infrastructure, and flood control, are fundamentally interrelated. The ongoing efforts to dredge Jakarta's rivers and to remove inappropriate uses from the riverbanks can have some short-term positive impacts on surface water quality.[31] Without a comprehensive and integrated regional water management strategy, however, the long-term effectiveness of the river dredging and normalization will be limited and likely compromised in the near future.

As discussed in Chapter 2, an integrated regional water management strategy enabled the city to function effectively in its early years. During the first (17th) century of the city's development as Batavia, harnessing the rivers to support the commercial and agricultural activities of the colonial port city and its hinterland, was the chief priority. Flooding was more an annoyance than the devastating event that it would become in the 20th century. Periodic water shortages (during the dry season) raised the specter that even with a relatively small population, an area served by 13 rivers still could face water shortages or the discomfort of living alongside stagnant waterways. So long as the urbanized area remained relatively compact, as it did throughout the 18th and early 19th centuries, the water management challenges remained the same. These involved securing potable water either from upstream intakes where the pollution levels were lower, or from groundwater; keeping the water channels clear by dredging regularly; and constructing canals outside the populated areas to divert water in support of agriculture and to mitigate the potential harmful effects of flooding in the colonial settlement.

Although the Dutch engineers were highly proficient at draining swampy soils and constructing barriers (dikes) to prevent seawater incursion back in the home country, the engineering applied in the East Indies was limited to canal construction. There were no windmills to support the pumping of unwanted waters, or much attention to strategies to maintain surface water resources for public consumption. By the late 19th century, the European population had come to rely heavily upon groundwater for human consumption, and with little investment in managing the water in the river system until some major flooding in the early 20th century.

Not until the construction in 1965 of the Jatiluhur Dam on the Citarum River 70 kilometers southeast of Jakarta at Purakarta did the city have access to a reservoir to provide a reliable source of potable surface water. Yet the use of surface canals rather than pipes to convey the reservoir water supply to the city, and the absence of water treatment facilities when it reached its urban destination, meant that Jatiluhur did not function effectively for long as a safe source of raw water unless it was treated and what little got treated was of minimal quality. This insufficient clean water system forced a large segment of Jakarta's growing population to rely on groundwater. By 2009, Jatiluhur provided just 60% of the supply for Jakarta's water company, with the rest coming from groundwater sources. A World Bank grant provided funds for low-income families to connect to one of Jakarta's two water providers. This never achieved its objective, and the water that they were supposed to get access to was of low quality. Based upon research derived from water samples along the 71 kilometers of the West Tarum Canal, it was highly polluted when it reached the city's treatment plants in Buaran and Pulogadang. The pollution occurred, according to the researchers, "because of the anthropogenic activities along the river bank, which is toilet activity ... domestic waste into the river, taking out trash and others."[32] This merely confirmed what had been widely known for years. The direct deposit of household waste water into the rivers is a condition that is readily apparent when one examines virtually all riverfront Jakarta kampungs where the waste pipes extend directly from the houses into the waterways, as illustrated in Figure 7.7.

Upon the recommendation of the Jakarta Water Regulatory Body (BR PAM), Governor Bowo announced late in 2009 plans to construct a water treatment facility at Jatiluhur and to connect it to Jakarta's water providers through a 72-kilometer pipeline. This US$312.5 million project, planned to be online by 2013, would have taken a major step toward the goal of every Jakarta household having access to safe drinking water. In the meantime, Jakarta's water supply companies would have to rely on water purchased from neighboring Tangerang to meet consumer demand.[33] The water treatment facility at the Jatiluhur and the piped connection to the Jakarta water companies never happened. The open canal, subject to the practices of the residents living alongside it who "feel that they have a right to do so since they have occupied the land for two decades," remained Jakarta's main raw water source from Jatiluhur and ensured that neither the quality nor the quantity the city needed would be any better.[34]

Dredging the waterways, removing squatter settlements from the banks of these rivers, and constructing an improved system for providing clean water all were preconditions to resolving long-standing deficiencies in Jakarta's water management system. Given Jakarta's history of having access to the best advice about water management, enactment of environmental protections, and specific projects to address the problems, the missing factor was the political will necessary for transforming plans into action. That quite rarely happened but actions did produce results. There is evidence, for example, that the long-awaited completion of the East Jakarta Flood Canal reduced flooding in 2010, according to the residents of the Cipinang Muara community. Before its completion, it was typical, as one resident noted, for houses to experience up to two meters of floodwater. Despite the seasonably heavy rains in February 2010, no flooding occurred in that portion of the canal already deep enough to accept water. Since the homes of residents in Cipinang Muara as well as in other riverfront community sit less than 20 meters from the new flood canal, the potential for future flooding remains until actions provide separation between the floodplain and the city's settlements. Introducing green infrastructure in support of these communities is a strategy that has not seen wide use in Jakarta.[35]

The World Bank-funded dredging between 2013 and 2019 reduced the intensity of duration of flooding. Overall, this US$172 million project repaired or reconstructed 52 kilometers of embankments, added new retention basins to handle approximately 80 cubic meters of water during the rainy season, improved pumping capacities in the Sentiong-Ancol area along the north coast, and

Figure 7.7 Drain pipes in kampung along Ciliwung in Central Jakarta.
Source: Photo by author.

extracted in excess of 3.4 million cubic meters of dredge and waste materials from the waterways. The result was that in the 34 most flood-prone ***kelurahans***, the average water levels during floods dropped from four feet to two feet, and tended to last just two days rather than the previous average of seven days. In addition, fewer areas within these flood-prone kelurahans experienced flooding, whereas prior to the dredging, the problem typically swamped the entire subdistrict. Clearly, the investment in river and canal management brought some mitigation of the severe circumstances experienced in previous floods.[36] Noteworthy was not that conditions were clearly better by the project's end in 2019, but that since 2013, there had not been the kind of devastating inundations that the city had experienced nearly every five years prior to that. What had not improved was the pollution of the surface water, the continued sinking of the city that left more of the city situated below sea level, and no systems in place to curb the pollution of surface waters that sustained the overreliance on groundwater.

Analysts stress the absence of, and need for, an integrated water management approach for the Greater Jakarta Area. Delinom (2008) focuses on groundwater management, which he approaches from the vantage point of two distinct areas of the megacity, areas of intense settlement (discharge areas), and areas where the groundwater system renews itself (recharge areas). Central to his model is to treat groundwater as a nonrenewable resource in order to underscore the potential (without constraints) of extracting more in the discharge area than any possible recharge process can accommodate. In the recharge areas, proactive steps are necessary, including land rehabilitation, reforestation, spring conservation, the introduction of artificial recharge mechanisms (such as injection wells), and expanded land area dedicated to recharging the groundwater. In the discharge areas, improved water treatment and waste management includes monitoring of water quality and quantity in wells, disincentives to groundwater extraction (groundwater taxes) and control of potential pollutants from residential and industrial uses (which are concentrated in discharge areas). All of these strategies require systematic implementation.[37]

The need for an "integrated urban water management" system in Jakarta was the theme of a World Bank presentation during Water Week in 2009. According to a model devised by Carlos Tucci, there were five interrelated issues to address through such a system. These include elimination of contamination in water sources, increased quantities of potable water, improved sanitation involving wastewater and solid waste removal, a strategy to manage stormwater, and addressing environmental degradation and the recurrence of negative health impacts of poor water quality for the urban population. One significant factor revealed in Tucci's analysis is that agricultural irrigation uses more than one half of the water supply serving Jakarta. At the heart of the problem for the future of the metropolitan area was the need to create an effective urban development regulatory process based upon a master plan for water management. Jakarta's water management plan needs to address water supply and distribution, waste collection and treatment, flood controls, and overall water and environment conservation. Overseeing the system involves strengthened management, a process of regular monitoring, capacity building to handle these responsibilities within the government, and a strong legal framework to implement the system. As Tucci noted, Jakarta needs "new urban development standards taking into account the sustainability on water issues," that protects water supply sources, improves the distribution system, upgrades the sanitation system, provides regulation for new development to ensure flood management for each basin, and an ongoing capacity to prevent sedimentation in river ways that inhibit drainage.[38]

Whether Jakarta has the legal and institutional capacity to address its present and future flood risk management challenges is not clear. Even with the delegation of authority to the local governments under Indonesia's decentralization system, Jakarta still lacks the authority to borrow funds to finance capital projects (such as river dredging, wastewater treatment, or to improve the water supply system).

Even more critical is that the localities surrounding Jakarta through which the rivers flow are independent authorities situated in two different provinces. This suggests the need for a regional management authority covering the Jakarta watershed. If the current river dredging, water supply expansion projects, and river clean efforts are to have a lasting impact, a water management organization with the power to sustain current improvements must be put in place. With an integrated water management system for Jakarta, even the challenges of climate change, sea-level rise, and the likely continued growth of the metropolitan population will not rule out the promise of an environment beneficial to all its residents.

Shaping the future in the water city

If an integrated urban water management system is the end goal, what steps are needed to get there in the short run? Every day, the Jakarta region continues to extend its urban footprint deeper into the surrounding green space where all of the city's surface water sources pass through the built environment. Out of necessity that built environment will continuously extract groundwater resources while doing little to replace it. Water experts agree that as long as this pattern of consumption continues, Jakarta's critical land subsidence problem will persist as the clean water supply decreases. Pollution prevents appropriate utilization of the seemingly abundant surface waters. Since the 1990s, the government has recognized the necessity to clean Jakarta's rivers but has avoided the appropriate and necessarily steps to correct this deficiency. Unless river restoration becomes a top policy priority, which means removing the ongoing sources of pollution coming largely from development within their boundaries, nothing will change. The dredging and the "normalization" of the rivers might curb some flood conditions but will do nothing to "restore" the quality and viability of the watershed to serve the city. To bring about river restoration in Jakarta requires support from the central government, collaboration with the surrounding communities, and engagement of business and civil society organizations. For Jakarta government, the biggest task involves expanding access to sanitary infrastructure, especially in low-income communities to reduce the practice of relying upon the rivers as their disposal site. Jakarta's government must vigorously enforce laws against industrial and commercial discharges into the rivers and streams along with a sustained public education campaign to support river revitalization. There is also a need to turn to innovative green infrastructure strategies that other cities have found helpful to manage stormwater and to restore water quality. Without immediate and sustained interventions to significantly improve river water quality, Jakarta's water crisis will not go away.

Restoring rivers to life occurs in other cities, and it was not so long ago that Jakarta's rivers, even with a century of neglect, represented an asset. To accomplish the restoration requires investment comparable to what now supports modern transit and the priority given over the past 50 years on highways and other infrastructure to encourage development of the built environment. Many global cities have found ways to reconnect the built environment with nature, but in the case of Jakarta, this will require much more than small increments to its green spaces. Phasing in a wastewater management system utilizing an updated and more comprehensive Kampung Improvement Program strategy would be a start. This involves adding additional high-density housing within the center city locations and installation of community-based wastewater management for existing kampungs as an alternative to the unfeasible comprehensive systems. Plans prepared by architects connected to several community-based organization show how to rebuild rather than destroy riverfront communities that have the potential to incorporate waste management technologies as well as flood-resistant designs. The use of green infrastructure to address issues of water pollution but also flood mitigation has been shown to work well in the case of Bishan Park in Singapore (as shown in Figure 7.8).

The argument to shift the focus from community clearance to community service upgrading is supported not just by the residents and their activist allies but by scientific research that suggests that Jakarta's primary challenge is not traffic congestions, or even flooding, but rather restoring water quality through management of pollution. The relationship between pollution of the surface waters

Figure 7.8 Bishan Park in the central area of Singapore represented a flood mitigation project coupled with the creation of a public park centered around the Kallang River freed some its encasement within a concrete ditch.

Source: Photo by author.

and flooding, overexploitation of groundwater, land subsidence, public health, as well as the quality of the megacity environment is determined directly by how water pollution is handled. An international team of researchers produced in 2019 what is undoubtedly the most comprehensive investigation of water quality in Jakarta, drawing upon existing studies examining land subsidence, water quality in several of the city's key rivers, and most importantly data from the 44 water quality stations managed by the government. These water quality stations provided data on all of Jakarta rivers, canals, and drains, including portions of the upstream, middle stream, and lower stream areas. While unable to pinpoint precisely how land use changes may have contributed to water quality, they did note that the significant decline in green areas during the period they collected the data undoubtedly affected the quality of their samples. Between 2007 and 2013, owing largely to commercial developments, the green areas of Jakarta dropped from 33,467 hectares (or 29% of the land area) to 10,108 hectares (or 9% of the land area). Reflecting on the role of retaining green areas to support the city water management system, they noted that "[f]lood prevention requires ample green space as does safeguarding ecological and human health. Green spaces also contribute to improving groundwater quality, flood prevention, improving air quality and decreasing urban temperatures."[39] Here again, the example of Bishan Park in high-density Singapore, as shown in Figure 7.9, demonstrates this approach as a response to the problem of water stress.

The key contributing factor to water stress in Jakarta, as their analysis shows, is the lack of comprehensive and effective wastewater treatment. Lacking a comprehensive wastewater treatment system, what Jakarta has instead, is a fragmented and ineffectual set of interventions that cannot combat pollution. The main source of wastewater management is through on-site technology, that is, individual septic tanks, serving approximately 8.4 million residents (or 71% of demand). Another 1.9 million residents connect to individual treatment plants (ITPs) that have accompanied new developments, some of which have secondary treatment before discharging into the river. A very small number of connections, covering approximately 200,000 residents, pipe directly to the Setiabudi Wastewater Treatment Plant. Another 1.3 million residents living in slum areas discharge untreated wastewater directly into rivers.[40]

As the researchers noted, and is evident when assessing the longer view of water policies in Jakarta, the government has chosen consistently not to actualize water improvement strategies set forth in plans. The Jakarta Mid-term Development Plan (2007–2017) includes an array of targets in the water sector, such as flood control systems and efforts to manage groundwater and wastewater treatment. Clearly, attention to flood control absorbed government resources throughout this period, although in fact the major initiatives of dredging and normalization, as well as improved protections along the northern coast, did not get started until midway during the plan, roughly concurrent with the construction of the Metropolitan Rapid Transit system and the light rail network. In terms of addressing water quality and the pervasive problem of polluted water, there were no comparable accomplishments to point to.

Figure 7.9 Allowing the Kallang River to migrate over a restored wetland provided the means to purify the water before it eventually found its way to the coastal area. This represented a case in nearby Singapore of the strategy of not having urbanization impact on the waterway.

Source: Photo by author.

Yet the researchers found some reasons to be optimistic. Their assessment of the water quality data indicated that there were, in some of the water quality stations, reductions in biochemical oxygen demand (BOD) and dissolved oxygen (DO) levels. Upstream stations often did better although some stations in the lower stream also evidenced improvements. Clearly, where human activities were most intense in the areas of the water quality stations, the results were poor. As they noted, the Jakarta government operates two septic tank wastewater stations that have the capacity of 300 cubic meters, and the use of these facilities may account for improvements in the readings between 2008 and 2014. When I lived in Jakarta in the mid-1990s, and hired a service to empty the septic tank in our rented house, I discovered that the reason for the very low cost of the service was that they simply took their load to the nearest stream and dumped it there. Back then, the operators did not need to pay a fee to a wastewater treatment facility. Obviously, enforcement of regulations against dumping, coupled with facilities to handle septic wastewater, can bring needed incremental improvements. Overall, however, the evidence of generally poor water quality, and the data on the high percentage of the population not served by sanitary infrastructure, shows that these incremental improvements are not enough.

What the research team concludes is that "[b)est practice dictates that improving water quality requires wastewater collection and connection to sewerage and ultimately to WWTP [Waste Water Treatment Plant]." They stress the centrality of managing water pollution to all components of the water management system to "eliminate over exploitation of ground water, prevent flood disasters, and use of ecological measures to clean urban water according to comprehensive urban development plans."[41] The indirect but equally significant impacts include making a more livable city environment, improved public health, tourism development, enhanced real estate values, and the economic outcomes likely when a megacity is able to promote itself as free of pollution. Certainly, that is a demonstrated outcome that the neighboring city Singapore takes advantage of as it extends its urban footprint further into the sea while at the same time boasting such clean waters that the gray water can be effectively treated to be consumed as "new water."[42]

Jakarta will not likely come close to restoring its surface water to the point that it will boast its own "new water" or perhaps even achieve a level of improvement comparable to what it was in colonial Batavia when the rivers served all of its needs. Nor will Jakarta completely wean itself from groundwater to end the problem of land subsidence and how it exacerbates flooding throughout the metropolis. But it is within the realm of doable to transform the vast network of surface waterway that provided the rationale of the city's development in the first place to be restored as an asset. This requires relying on science, sound engineering, creative design efforts, social policy that recognizes the necessity of ensuring that all of its citizens have their basic needs met within their means, and ongoing planning and management processes that recreate and sustain livable waters. And it will require an investment by the government, in concert with private interests that are likely to benefit, comparable to the investment already committed to transit modernization. Yet in the long run, the returns from the investment, when viewed through the broad perspective of recreating a livable city, will be just as great. The present and future citizens of Jakarta deserve it.

A change of direction in the approach to flood control and water management in Jakarta seems seriously overdue. It must begin with recognition that the city's rivers are a precious asset, an asset that can make Jakarta a healthy and intensely livable city or be a continuing source of degradation. The change begins by embracing the concept that rivers must have space to ebb and flow, to be protected and nourished by wetlands not impacted by urban development, and accessible to Jakarta's population in positive ways. There is a vast array of models to support restoration and maintenance of healthy rivers that have overcome the levels of degradation comparable to the conditions in Jakarta. As long as it is possible and acceptable to openly discharge wastes in Jakarta's waterways, and no program to redress what has already been done, the goal of restoration cannot be accomplished. There are a variety of small but essential first steps that can begin the process. Removing direct discharge through vigorous and community-based enforcement throughout the watershed is essential. Reestablishing "green" and not just "concrete" barriers along the rivers is important. Reconstructing wetlands recharge areas, most likely upstream, can address water quality. These areas can also be constructed to allow the public to access to them for recreational use, contribute to restoring natural processes, and offer the public a different experience with water. Residents can reconnect with nature and in the process gain greater appreciation of environmental stewardship.

The values of and prescriptions for the biophilic city, as set forth by Beatley (2011), are a necessary starting point for reimagining and restoring the ecological vitality of Jakarta. Without specifically referring to Jakarta (although he certainly could be), Beatley observed that

> the presence of a river, or confluence of several rivers, is the major historical reason a city exists where it does, but too often the river's edge habitats have been degraded or destroyed, its water quality and other aquatic values have been compromised, and the city has physically turned its back to river.[43]

In many cities that allowed these conditions to persist, ambitious river restorations have succeeded in reconnecting cities to their ecological heritage and, in the process, have become greater places in every respect. Beatley contends that through concerted interventions, "the most degraded urban rivers and streams" can be restored, "reconnecting urban populations to this natural hydrology."[44] In the case of Jakarta, the benefits go beyond the aesthetics of connecting with nature. It is a prerequisite to a healthy life for all of its millions of residents. As daunting as the task may seem in the case of Jakarta, other cities have shown that it is both doable and worthwhile.

Notes

1 Khafi, Kharishar (2020) "Not Ordinary Rain: Worst Rainfall in Over Decade Causes Massive Floods in Jakarta," *Jakarta Post*, January 1.
2 Atika, Sausan (2020) "Flooding Becomes Politicized as Anies Faces Criticism: Jakarta, Lebak Floods Claimed Over 60 Lives," *Jakarta Post*, January 16, p. 5.
3 Interview with Jakarta Deputy Governor, January 14, 2020.
4 Ghalina, Ghina and Tehusijarana, Karina M. (2019) "Borneo, Sulawesi Top Capital Candidates," *Jakarta Post*, May 2.
5 Sapiie, Marguerite Afra (2019) "Jokowi Wants to Move Capital Out of Java," *Jakarta Post*, April 29.
6 Atika, Sausan (2019) "Jakarta to Start River Naturalization Projects," *Jakarta Post*, May 4.
7 "Cure to Sinking Jakarta," Editorial, *Jakarta Post*, July 30, 2019.
8 Wijaya, Callistasia Anggun (2016) "Development of Jakarta Giant Seawall Crucial to Anticipate Land Drown, Minister Says," *Jakarta Post*, June 6.
9 Ibid.
10 "Seawall in Stalemate as Firms Await Government Decision," *Jakarta Post*, December 27, 2016; "Seawall Project to Be Completed This Year: Minister," *Jakarta Post*, March

9, 2017; "Seawall Construction in Pasar Ikan, Kali Blencong Nearly Finished: Official," *Jakarta Post*, November 23, 2018.

11 "How the People of Jakarta View Their Changing City," *Jakarta Post*, June 22, 1996.

12 "Mass Rapid Transit System Required for the Capital," *Jakarta Post*, February 12, 1990; "How the People of Jakarta View Their Changing City," *Jakarta Post*, June 22, 1996; "Bimantara Boss Joins US$1.5b MRT Project," *Jakarta Post*, September 9, 1996; "Subway Will Start in June," *Jakarta Post*, March 6, 1997.

13 Iswara, Made Anthony (2019) "Southeast Asia's First High-Speed Train Ready for Operation by 2021," *Jakarta Post*, May 16.

14 Abidin, Hasanuddin Z, Andreas, Heri, Gumilar, Irwan, Fukuda, Yoichi, Pohan, Yusuf E., and Deguchi, T. (2011) "Land Subsidence of Jakarta (Indonesia) and Its Relation With Urban Development," *Natural Hazards*, 59 (3): 1753–1771; Ng, A.H-M., Ge, L., Li, X., Abidin, H.Z., Andreas, H., and Zhang, K (2012) "Mapping Land Subsidence in Jakarta, Indonesia Using Persistent Scatterer interferometry (PSI) Technique with ALOS PALSAR," *International Journal of Applied Earth Observation and Geoinformation*, 18 (1): 232–242.

15 Colven, Emma (2020) "Subterranean Infrastructures in a Sinking City: The Politics of Visibility in Jakarta," *Critical Asian Studies*, 53 (3): 321–322.

16 Ibid., pp. 324–325.

17 Caljouw, Mark, Nas, Peter J.M., and Pratiwo (2005) "Flooding in Jakarta: Towards a Blue City with Improved Water Management," *Bijdragen tot de Taal-, Land-en Volkenkunde*, 161 (4): 464.

18 Bakker, Karen (2007) "Trickle Down? Private Sector Participation and the Pro-Poor Water Supply Debate in Jakarta, Indonesia," *Geoforum*, 5: 857.

19 Sedlar, Frank (2016) "Inundated Infrastructure: Jakarta's Failing Hydraulic Infrastructure," *Michigan Journal of Sustainability*, 4 (Summer): 1–11.

20 Manan, Rustam Hakim (2016) "Policy Analysis of Urban Green Open Space Management in Jakarta City, Indonesia," *International Journal of Engineering and Technology* 5(4): 241–248.

21 Abeyasekere, Susan (1989) *Jakarta: A History*, Revised Edition. New York: Oxford, p. 226.

22 Putri, Parthiwi Widyatmi (2019) "Sanitizing Jakarta: Decolonizing Planning and Kampung Imaginary," *Planning Perspectives*, 34 (5): 818; Bakker (2007), pp. 863–865.

23 Rukmana, Deden (2013) "Jakarta Annual Flooding in January 2013," *Indonesia Urban Studies*. http://indonesiaurbanstudies.blogspot.com/2013/05/jakarta-annual-flooding-in-january-2013-html.

24 Hidayatno, Akhmad, Dinianyadharani, Aninditha Kemala, and Sutriso, Aziz (2017) "Scenario Analysis of the Jakarta Coastal Defense Strategy: Sustainable Indicators Impact Assessment," *International Journal of Innovation and Sustainable Development*, 11 (1): 37–52.

25 Douglass, Mike (2010) "Globalization, Mega-Projects and the Environment: Urban Form and Water in Jakarta," *Environment and Urbanization ASIA*, 1 (1): 47.

26 Ibid., p. 46.

27 Steinberg, Florian (2007) "Jakarta: Environmental Problems and Sustainability," *Habitat International*, 31: 354–365; Argo, Teti A. (1999) *Thirsty Downstream: The Provision of Clean Water in Jakarta*. PhD dissertation, University of British Columbia, Vancouver, Canada; Bakker, op. cit.

28 USAID, WASH (2019) Accessed January 29, 2020: www.usaid.gov/actingon the call/stories/indonesia-wash2019.

29 Putri (2019), p. 805.
30 Ibid., p. 818.
31 World Bank (2019) *Implementation Completion and Results Report*, IBRD-8210, Jakarta Emergency Dredging Initiative, August 23. http://documents.worldbank.org/curated/pt/153081567169469254/pdf/Indonesia-Jakarta-Urgent-Flood-Mitigation-Project.pdf.
32 Siabutar, Noni Valeria, Hartono, Djoko M., Soesilo, Tri Edhi Budhi, and Hutapea, Reynold C., "The Quality of Raw Water for Drinking Water Unit in Jakarta-Indonesia," International Conference on Chemistry, Chemical Process and Engineering (IC3PE) 2017, in *AIP Conference Proceedings* 1823, 020067-1020067-9; doi: 10.1063/1.4978140.
33 *Jakarta Post*, November 5 and 25, 2009.
34 Siabutar et al., op. cit.
35 *Jakarta Post*, February 17 and July 5, 2010.
36 World Bank (2017) *Jakarta Urgent Flood Mitigation Project Overview*. Accessed August 12, 2020: www.worldbank.org/P111034/jakarta-urgent-flood-mitigation-project/en/overview.
37 Delinom, Robert M. (2007) "Groundwater Management Issues in Greater Jakarta Area, Indonesia," Proceedings of International Workshop on Integrated Watershed Management for Sustainable Water Use in a Tropical Region, JSPS-DGHE Joint Research Project, Tsubuka University, October. Bulletin TERC University of Tsubuka, No. 8, Supplement (2): 40–54.
38 Tucci, Carlos E.M. (2009) "Integrated Urban Water Management in Jakarta," World Bank – Water Week, February 17–19; Tucci, Carlos E.M. (2009) *Integrated Urban Water Management in Large Cities: A Practical Tool Assessing Key Water Management Issues in the Large Cities of the Developing World*. Washington: World Bank.
39 Luo, Pingping, Kang, Shuxin, Apip, Meimei Zhou, Lyu, Jiqiang, Aisyah, Siti, Binaya, Mistra, Ram, Krishna Regmi, and Nover, Daniel (2019) "Water Quality Trend Assessment in Jakarta: A Rapidly Growing Asian Megacity," *PLoS ONE*, 14 (7): 5. Accessed April 10, 2020: http://doi.org/10.1371/journal.pone.0219009.
40 Ibid., p. 6.
41 Ibid., p. 13.
42 Tan, Thai Pin and Rawat, Stuti (2018) "NEWater in Singapore," *Global Water Forum*, January. Accessed April 22, 2020: http://globalwaterforum.org/20218/01/15/newater-in-singapore.
43 Beatley, Timothy (2011) *Biophilic Cities: Integrating Nature Into Urban Design and Planning*. Washington, DC: Island Press, pp. 95–98.
44 Ibid.

Index

9780367774271